Computer Speech Technology

For a complete listing of the *Artech House Signal Processing Library,*
turn to the back of this book.

Computer Speech Technology

Robert D. Rodman

Artech House
Boston • London

Library of Congress Cataloging-in-Publication Data
Rodman, Robert D.
    Computer speech technology / Robert D. Rodman.
        p.   cm.
    Includes bibliographical references and index.
    ISBN 0-89006-297-8 (alk. paper)
    1. Automatic speech recognition.   2. Signal processing.
I. Title. II. Series.
TK7895.S65R63   1999
006.4'5—dc21   98-51804
                                                                            CIP

British Library Cataloguing in Publication Data
Rodman, Robert D.
    Computer speech technology
    1. Speech processing systems
    I. Title
    006.4'54

    ISBN 0-89006-297-8

Cover design by Lynda Fishbourne

© 1999 ARTECH HOUSE, INC.
685 Canton Street
Norwood, MA 02062

International Standard Book Number: 0-89006-297-8
Library of Congress Catalog Card Number: 98-51804

10 9 8 7 6 5 4 3 2 1

To the memory of Cecelia

Contents

Preface

Complaining is how this book got written. My complaints date back to the early 1980s, when I reproached designers and builders of speech recognition systems for not taking fully into account the nature of language and language usage. I finally stopped cursing the darkness and lit a candle. This book is the result.

Two motifs distinguish this book from other books on the subject. One is the continual attention to the nature of human language and its impact on the science and methodologies of computer speech technology. The other is a nod in the direction of Aldous Huxley, who once wrote:

> I admit that mathematical science is a good thing. But excessive devotion to it is a bad thing.

Computer Speech Technology presents all the concepts of speech processing in ordinary English through use of illustrations and analogies accessible to most readers. Where necessary, complex mathematical matters such as Fourier transforms are explained in plain language, with sufficient mathematical formulae for the reader to gain an appreciation of the depth of the field, without being overwhelmed by it. References to more technical material are given throughout the book for readers who require additional technological depth.

This book, then, is written for individuals with an interest in computer speech technology other than a deep mathematical one. This includes speech technology sales personnel, consultants, managers, marketing professionals, and end users. Psychologists, industrial engineers, human-computer interface practitioners, speech and hearing professionals, and health care professionals—all of whom may come in contact with speech technology—should find this book of interest. Interest extends to the computer hobbyist—much computer software bears on speech processing—as well as to individuals who are simply curious about how we get computers to talk and listen and identify people by their voices.

The book is divided into three parts: fundamentals, uses, and applications. Fundamentals are covered in two chapters. Chapter 1 is about human speech production and the science of phonetics. Chapter 2 takes up the basic concepts of computer speech processing such as sampling, quantization, waveform coding, and parametric coding.

The middle chapters discuss the various areas in which speech processing is used. Chapter 3 is about computer speech recognition, the most prolific use of speech processing. It is the book's longest chapter. All types of speech recognition systems, and their underlying technology, are discussed and illustrated. There is sufficient discussion of error analysis and performance evaluation to allow potential users of speech recognition technology to make measured purchasing decisions.

Chapter 4 is on speech synthesis, or getting computers to talk like you and me. How do computers speak, how do they know what to say, and how well do humans understand them, are some of the questions addressed in this chapter.

Chapter 5 is on speaker recognition, or "voiceprints." The fundamental question is can a computer identify you by your voice alone, the way you do when your parent, child, or spouse calls on the telephone. After the requisite technical background we'll be able to discuss applications such as identifying the 1996 Olympic Park bomber.

Also in Chapter 5 are two interesting, though lesser known uses of speech processing. One is language identification. Can a computer learn to identify the language being spoken, whether English, Japanese, or French? The other is lip synching. Given any freely chosen speech as input, can a computer shape the lips and tongue of a graphical talking head as if it were speaking the speech, and do it in perfect synchronization? The answers to these questions are found in this book.

The final three chapters are concerned with applications of the technologies discussed in the middle three chapters. Chapter 6 considers the manifold applications of speech recognition such as speech controlled robots as aids to the disabled, and communications with computers in hostile environments such as space (ever try typing in a space suit?).

Chapter 7 discusses applications of speech synthesis, starting with the once ubiquitous "at the tone, the time will be. . . ."—older readers will remember—

and including automatic reading machines, spoken instructions in blacked out environments, and having your e-mail read to you by your computer over the telephone.

Chapter 8 takes up applications of speaker recognition. We'll consider the sinister, such as identifying persons leaving recorded death threats; the serious, such as preventing cellular phone fraud, a multi-billion dollar rip-off; and the silly, determining whether the voice of the mother of Ferdinand the Bull in the 1930s movie cartoon belonged to Walt Disney.

I have striven throughout the book to maintain a style that is both light and lucid. These are the pinnacles of art, whether writing, performing a Mozart sonata, or baking a soufflé.

In writing I've tried to keep you, the reader, in mind. Occasionally I speak to you on a personal level, just as I would if we sat in a pleasant place and discussed the topic at hand over refreshments. Thus the pronouns *you* and *I* appear more frequently in this book than in most others of its kind.

I am deeply grateful to the individuals who helped me at various stages of producing this book, from its inception to its printing. If I've forgotten to mention you, dinner's on me. What the heck, if I remembered to mention you, dinner's on me. Many thanks to Don Bitzer, Daryle Gardner-Bonneau, Rick Klevans, Ron Mace, Dave McAllister, Zack Rodman, Mark White, and the many students who used early versions of the manuscript as their text book in class, and offered their suggestions for improvement. I want to especially thank my wife Helen for her unflagging support, both moral and technical, during the several years of research and writing. Finally, I dedicate this book to the memory of Cecelia. She knows who she is. She always did.

1

About Speech

In a continuous discourse there is no separation between the words except where we pause to take breath, or for emphasis: the words of a sentence are run together exactly in the same way as the syllables of a word are.

Henry Sweet, *A Primer of Phonetics,* 1890

1.1 Introduction

When we hear speech in our own language, we hear individual words and sounds. This truth is easily proven: You can write down speech using discrete letters with spaces between words. When we hear speech in a language that we do not know, we cannot do this. The words and sounds all seem to run together in a continuous, indivisible stream. Which of these accounts of speech is true? Is speech discrete, or is it continuous?

Surprisingly, the answer is both. On a purely physical level, speech is continuous, "except where we pause to take breath," as Professor Henry Sweet (the prototype for *My Fair Lady's* Henry Higgins) would have it. On a psychological level, speech is perceived as composed of discrete sounds and groups of discrete sounds, the former corresponding more or less to the letters of the alphabet and the latter to words of the language.

Humans are capable of dividing physically continuous speech signal into discrete units because of their linguistic acumen. Mastering language is arguably the most prodigious single intellectual accomplishment of a person's life. An unfathomable amount of knowledge is required to speak and understand a language with native fluency. Much of this knowledge is brought to bear in processing the speech signal.

1

This dual nature of speech—discrete or continuous—is what makes computer processing of speech so challenging. Consider the digital computer, capable of only incompletely representing a continuous signal, devoid of linguistic knowledge except for the trifling amount provided by us humans. The computer is little better off than you in a Parisian taxi cab (assuming that you do not know French).

Chapter 2 examines what can be done to help the computer deal with speech. This chapter looks at speech from the human perspective. This is the true perspective in the sense that speech and language are unique to the human species.[1]

1.2 How Speech Is Produced

It could be argued that speech science is 2,500 years old. Evidence for this claim is found in the writings of the great Greek physician Hippocrates (460–377 B.C.):

> The voice is articulated by the lips and the tongue. . . . Man speaks by means of the air which he inhales into his entire body and particularly into the body cavities. When the air is expelled through the empty space it produces a sound, because of the resonances in the skull. The tongue articulates by its strokes; it gathers the air in the throat and pushes it against the palate and the teeth, thereby giving the sound a definite shape. If the tongue would not articulate each time, by means of its strokes, man would not speak clearly and would only be able to produce a few simple sounds.

This description of speech is remarkably apt. The concept is correct in its essentials, though modern linguists have filled in the details, many of which are discussed in this chapter.

1.2.1 The Vocal Tract

Before he invented the telephone, Alexander Graham Bell attempted to build a model of the human vocal tract, a talking head, if you will. The paucity of materials available to science in the mid-nineteenth century made the task a formidable one, and it is to Bell's credit that his invention produced a few speech

---

1. Parrots and their like imitate speech but lack language. Other animals possess communication systems, but none meets the criteria of human language, including primates that use sign language. For a complete discussion of this subject, see [1].

sounds. One of the easily obtainable components of the talking head were bellows, needed because the device was air-driven. These correspond to human lungs, which, during the production of speech, force air into the windpipe, through the vocal cords, up the pharynx, and out the nose or mouth. Human speech production is an air-driven process, powered by the lungs. (See Figure 1.1 for an anatomical view.)

The sounds that are emitted during speech depend on what happens to the air on its journey from the lungs. The first obstacle is the vocal cords. These are a pair of muscular folds that span the larynx and are capable of various degrees of closure. When they are wide apart, air passes through the larynx without disturbing them. A speech sound produced with this type of air flow is said to be **voiceless.** The initial sounds of *seal, feel,* and *peal,* represented by *s, f,* and *p,* are voiceless sounds.

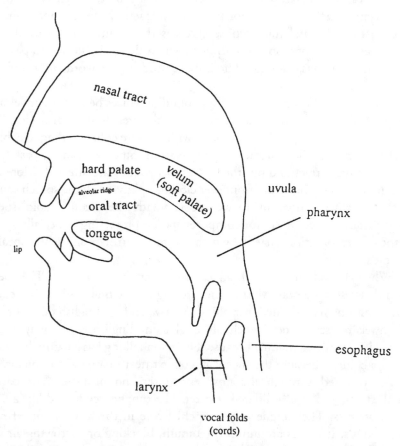

Figure 1.1 The human vocal tract.

When the vocal cords are closed, air may still pass but the vocal cords are forced to vibrate. You may sense these vibrations by cupping your hands over your ears and saying z-z-z-z-z-z-z-z-z-z-z-z-z-z-z-z. You will not hear the same thing if you say s-s-s-s-s-s-s-s-s-s-s-s-s because *s* represents a voiceless sound. Sounds produced when the vocal cords are together are said to be **voiced**. The initial sounds of *zeal, veal,* and *real* are voiced sounds.

The vocal cords may be partially closed. The sounds produced are described as either **murmured** or **creaky,** depending on the shape of the opening. Such sounds do not play a significant role in English, though they do in other languages. They are important because they are variants of voiced sounds that may affect computer speech processing.

The next station for air passing out of the lungs during speech occurs at the **soft palate** or **velum.** To locate the velum, place a thumb against the roof of your mouth and slide it inward toward the throat. You will first feel the hard, bony part of your palate and then the soft, fleshy part, which is where the velum begins. The velum is a door to the nasal tract. When it is open, air may pass both into the oral tract, whose egress is the mouth, or the nasal tract, whose egress is the nose. Speech sounds in which the air is allowed to pass into the nasal tract are **nasal** sounds. The initial sounds of the words *meal* and *Neil* are nasal sounds.

Some people have nasal or twangy sounding voices because their velum is such that it does not always make a complete closure, allowing air to leak into the nasal tract during speech. In a person with a severe cold, swelling may block the nasal tract regardless of the velum's position. Since air cannot pass freely through the nasal tract and out the nose, nasal sounds are affected. Moreover, mucus may prevent the velum from ever closing tightly, so all the speech sounds have that nasally distorted quality we associate with a head cold. "Cold" speech has significant consequences for computer speech processing, especially speech and speaker recognition systems, which are more error-prone when speakers have colds.

When the velum is closed, air cannot enter the nasal tract. The speech sounds that result are **oral.** Most sounds in English are oral, with the exception of those consonants indicated in spelling by *m, n,* and *ng.* English does not have nasal vowels per se, as does French, although some English vowels may be pronounced in a nasalized manner because the person talking has a naturally twangy voice or a cold, or because they occur in context next to a nasal consonant.

Air diverted through the nose encounters no obstacles. Air passing through the oral tract is a different matter. The tongue, teeth, and lips are potential obstacles. The tongue plays the chief role in the articulation of most speech sounds. It moves around in the mouth, blocking or narrowing air flow between itself and the velum, hard palate, or teeth. The lips may be open, closed, puckered, or moved against the teeth. The different positions of these articulators result in different speech sounds.

The tongue, teeth, and lips may work together to produce the desired sound. For example, the placement of the tongue behind the upper teeth produces the *t, d,* and *n* sounds of *tip, dip,* and *nip,* depending on other factors such as voicing and nasalization. The tongue forces air against the upper teeth or hard palate when the *s* and *z* sounds of *sip* and *zip* are produced.

Try saying the word *fie* while prolonging the *f.* You should feel your lower lip moving under your upper teeth, providing just the right obstruction for the sound *f.* With voicing, that is, when your vocal cords are vibrating, you will hear a *v* sound instead, as in *vie.*

Momentarily blocking the outgoing air with the lips alone is what produces the initial sounds of *potato, bagatelle,* and *machine.* Again, the differences among *p, b,* and *m* sounds have to do with voicing and nasalization. Try saying *the mamas and the papas.* You will feel your lips close and open four times, twice for the *m*'s and twice for the *p*'s. If you pinch your nostrils while carrying out this experiment you will notice that the sound of *mamas* changes whereas the pronunciation of *papas* is barely affected. This is because *m* is a nasal sound, but *p* is an oral sound.

Figure 1.1 is a schematic view of the vocal tract. A few of the labeled items have not yet been discussed. The **uvula**, for example, is at the end of the velum. It's the little flap that you see hanging down when you look in the mirror, open your mouth wide, and say "aaaahhhhhhhh." The vowel sound so awkwardly written as "aaaahhhhhhhh" forces you to lower your tongue out of the way to reveal the uvula (as well as the back of your throat) for examination. The **alveolar ridge** is a small protrusion of the hard palate just behind the upper front teeth. It's the part of your mouth you burn when you are too hungry to let the pizza cool.

1.2.2 Articulatory Phonetics

Phonetics is the scientific study of the speech sounds of human language. **Articulatory phonetics** is the branch of phonetics that deals with the physical process of human speech production. An understanding of how speech sounds are produced will help us, among other benefits, learn a foreign language, enunciate our own language more clearly, and design computer systems capable of producing and recognizing speech.

The vocal cords, tongue, velum, and lips are the moveable parts of the vocal tract that work together to produce the sounds that make up speech. The teeth, alveolar ridge, hard palate, and velum (again) play a passive role in speech production. All these parts of the speech-producing anatomy are called **articulators.**

Speech sounds occur in two types: consonants and vowels. Speech is mostly vowel sounds with intervening consonant sounds. Although this fact is not obvious to the untrained ear, instrumental measurements prove it to be true.

We first learn about consonants and vowels through the alphabet, but it must be emphasized that the alphabet is merely a way, and not a very precise way, of referring to speech sounds. Speech is primary over writing, and it is our main concern in this book.[2] For now we will make use of the alphabet because of its familiarity.

The consonant-vowel dichotomy is reflected in the alphabet. Most of us learned that the vowels are *a, e, i, o,* and *u,* and sometimes *y.* What this really means is that the letters *a, e, i, o, u,* and *y* are used to represent vowel sounds. We will see that English has far more than five or six vowel sounds. There is some equivocation regarding *y* because it signifies a vowel sound in *sky,* a different vowel in *silly,* and yet a third vowel in *system,* but a consonant in *year* and *beyond.*

Consonant sounds are represented by the other letters. How many consonant sounds[3] does English have? The answer, if we believe in the alphabet, is twenty-six minus five, and sometimes minus six—but should we believe the alphabet? Problems with *y* aside, we find *c* representing two consonants, one in *city,* another in *coat.* (Say these out loud if you're not convinced.) It takes two consonant letters, *t* and *h,* to make up the single consonant that begins *think,* and *x* in a word like *fox* actually represents two consonants, as if the word were spelled *foks.*

It is vital that persons interested in speech processing have thorough knowledge of the speech sounds of English (or whatever language they are involved in), which are not clearly represented by letters of the English alphabet. Sections 1.2.2.1 and 1.2.2.2 take a closer look at consonant and vowel sounds.

1.2.2.1 Consonants

Speech sounds pronounced with obstructions in which the articulators come close to each other, or touch, are called **consonants.** Vowel sounds are articulated without any such obstructions.

The sound of a consonant is determined by three factors: the location in the vocal tract of the obstruction, or obstructions, called **place of articulation;**

---

2. We say that speech is primary because all humans without severe disabilities or social deprivation learn to speak their native language without formal instruction. Few humans learn to write without formal instruction. Many perfectly functional and intelligent people within their own society never learn to read or write. Spoken language is a genetically endowed ability similar to bipedal locomotion. Writing is learned in much the same way as arithmetic.

3. To avoid the verbose terms *vowel sounds* or *consonant sounds,* I may use simply *vowel* or *consonant.* Unless I specifically refer to a symbol, the terms *vowel* and *consonant* mean "vowel sound" and "consonant sound."

the relative position and activity of the articulators during the obstruction, called **manner of articulation;** and the state of the vocal cords, whether apart for voiceless sounds, or together for voiced sounds, called **voicing.**

There are eight places of articulation important in a discussion of English. These do not cover all of the possibilities realized in the world's languages. Our discussion would be incomplete for French, Arabic, and Hebrew. Figure 1.2 outlines the places of articulation in the mouth.

Starting at the front of the mouth and working back toward the throat, the two lips may obstruct the passage of air by momentarily coming together. This place of articulation is called **bilabial.** The bilabial consonants of English are represented by the letters *p, b,* and *m.*

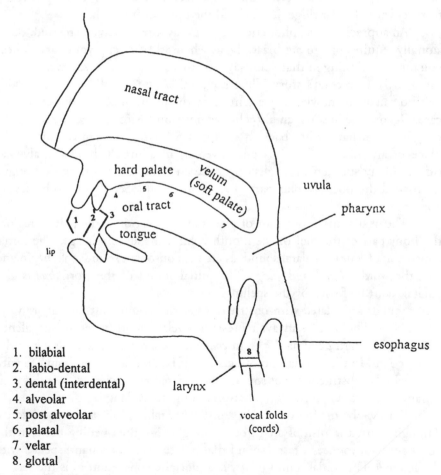

1. bilabial
2. labio-dental
3. dental (interdental)
4. alveolar
5. post alveolar
6. palatal
7. velar
8. glottal

Figure 1.2 Places of articulation: (1) bilabial, (2) labio-dental, (3) dental, (4) alveolar, (5) post-alveolar, (6) palatal, (7) velar, and (8) glottal.

Next is **labio-dental**. The obstruction is created by raising the lower lip toward the upper teeth. The letters *f* and *v* represent the labio-dental consonants of English.

Dental is the next place of articulation. The obstruction is created by the tongue tip and the teeth. Sounds spelled with a *th* as in *thy* and *thigh* are dental consonants.

When the front of the tongue is raised toward the alveolar ridge to create an obstruction, the place of articulation is called **alveolar.** Sounds spelled with *t, d, n, s, z,* and *l,* among others, represent alveolar consonants. If you are unsure of where your alveolar ridge is, and you don't like the idea of finding it by burning it on hot pizza, try saying *pizza.* The consonant represented by the *zz* (an Italian spelling) is alveolar. You should feel the front of your tongue rising to meet your alveolar ridge, just behind the upper front teeth.

An appreciation of phonetics, as you may already have perceived, occasionally requires one to act bizarrely, as when asked to cup your ears to hear voicing. Continuing in that vein, say *s-s-s-s-s-s-s-s-s-s-s* as if you were imitating a snake in a child's story. Then say *sh-sh-sh-sh-sh-sh-sh* to the noisy kids behind you at the movies. Pay attention to the position of the tongue in each case. (Ignore the lips, which also change position.) Most likely, you'll feel the tongue somewhat farther back in the mouth for the *sh* than for the *s.* The place of articulation for *sh* is post-alveolar, meaning "behind the alveolar ridge." Many speakers of English produce an *r*-sound by curling the tongue up toward the **post-alveolar** part of the hard palate. See if you do by saying *r-r-r-r-r-r-r-r-r-r.*

The next stop on our tour of places of articulation is the **palatal** region, the bony part of the roof of the mouth behind the alveolar ridge. The initial consonant of *year* is a palatal sound, as are the consonant sounds that begin and end the words *church* and *judge.* The initial sound of the word *hue* is also palatal; don't be fooled by the spelling.

Sounds articulated with obstructions at the velum (soft palate) are **velar** consonants. The letter *k* always represents a velar sound (when it isn't silent). The so-called hard *c* and hard *g,* as in *coat* and *goat,* represent sounds with velar obstructions. In the German pronunciation of Bach, the *ch* is pronounced with a partial velar obstruction that sounds like a throat being cleared. It is this consonant that gives German its reputation as a "guttural language."

Finally, obstruction of certain consonants takes place at the vocal cords. Though apart, the rush of air through the **glottis**—the opening between the vocal cords—produces a raspy sound that is actually a consonant, spelled with the letter *h.* The initial sound of *hat* is a **glottalic** consonant, as is the second consonant of *behind.* Try the usual trick of prolonging the first sound in *hat* to hear this sound.

As well, the glottis may be shut so tightly that air cannot pass at all, momentarily. This happens when you cough, for example. It also occurs during speech, although you may be unaware of it. In the expression of consternation spelled *uh oh,* there is a consonant articulated at the beginning of the *uh* and the *oh* that results from a total closure of the glottis. In many dialects of English, for example, the so-called Cockney dialect of British English, the consonant represented by *tt* in the word *bottle* is actually pronounced with a stoppage at the glottis rather than at the alveolar ridge. Thus, the vocal cords have the dual function of controlling voicing and being the place of articulation for certain consonants.

To summarize, eight places of articulation are significant in the pronunciation of English consonants. They are listed in Figure 1.2 and indexed to the schematic vocal tract.

Manner of articulation is the term for what the articulators do at the site of the obstruction.

Complete blockage is one possibility. Sounds articulated in this manner are called **stop consonants,** or simply **stops.** The initial sounds of *ball, doll, and gall,* and the second sounds of *spill, still, and skill,* are stops. The momentary, total obstruction of the airstream, followed by its sudden resumption, gives the stop its characteristic sound. A somewhat different sounding stop occurs as the initial sounds of *pill, till,* and *kill.* These stops are held long enough to allow air pressure to build up behind the obstruction. Their release is accompanied by an audible puff of air called **aspiration.** The difference between the *p* of *pill* and the *p* of *spill* is subtle and not always easy to hear, but you can detect the difference by putting your hand one inch from the front of your mouth and speaking the two words. You should feel the aspiration—the puff of air—when saying *pill* but not when saying *spill.* The effect is even more subtle for the *t* and *k* sound, but it is nonetheless present in most English speakers' pronunciations.

In a **nasal stop,** air is blocked in the mouth but allowed to flow out the nose. The final sounds of *beam, bean,* and *bing* are all nasal stops. In *beam* the stop is at the lips, so it is a **bilabial nasal stop.** You can prolong the *m* as in beam-m-m-m-m-m-m and, though the lips are tightly shut, air escapes via the nose. Do the same with *bean-n-n-n-n-n-n* and *bing-ng-ng-ng-ng-ng-ng-ng.* You will feel an alveolar stop in the former, and a velar stop in the latter, as the airstream moves through the nasal tract and out the nose. (This is an excellent way to get a feel for stop articulation since nasal stop consonants can be prolonged. If I asked you to draw out the final sound of *up,* which is a bilabial oral stop, you couldn't do it for long because the air would have no place to go.)

When two articulators are very close, but not touching, the airstream is partially obstructed. The resulting turbulence produces a sound called a

fricative. Sounds represented in English spelling by *f, v, th, s, z, sh,* and *h* are fricatives. All these sounds can be prolonged because the closure is incomplete. Thus, you can practice feeling where the place of articulation is by going *f-f-f-f-f-f-f,* or *th-th-th-th-th.*

When the approach of the articulators is not close enough to cause turbulence, the sound produced is an **approximant.** In English, the approximants are spelled with the letters *y, w, r,* and *l,* as in *yet, wet, ret,* and *let.*

It is possible for two manners of articulation to be combined. This occurs in English in the consonants beginning the words *chump* and *jump.* The *ch* sound is the alveolar stop *t* run together with the palatal fricative *sh.* This combination is called an **affricate,** and it is yet another common manner of articulation. You can gain a sense of what's going on here by saying *white shoes* without pausing between the words. It will be nearly indistinguishable from the two words *why choose.* The *t* of *white* combines with the *sh* of *shoes* to sound like the *ch* of *choose.* The *j* sound is the voiced counterpart of the voiceless *ch.* It is both the initial and final consonant in *judge,* the *j* and the *dge* being the two most common spellings.

The last manner of articulation of interest is a **flap.** It occurs in many varieties of American English as the middle consonant in words like *rider, writer, latter,* and *ladder* spoken in smoothly flowing speech. Speakers may be unaware that they pronounce something other than a *d* or *t* in these words. The flap is articulated when the tongue makes a single, fleeting tap against the alveolar ridge. Say any of these words rapidly while concentrating on what your tongue does.

To summarize, there are six manners of articulation significant for English. They are stop, nasal stop, fricative, approximant, affricate, and flap.

Consonants in English, then, may be described in terms of three parameters: place of articulation, manner of articulation, and voicing. In Table 1.1, the underlined letter(s) represents the consonant of interest. The consonants are voiced unless they are in a row marked "vcless." The table is somewhat simplified in that there are other consonant sounds that occur in English that we have not discussed. The interested reader may refer to a number of books on English phonetics for a more complete description, such as [2].

If you are willing to use technical jargon, you can describe consonants directly without resorting to expressions like "the sound represented by the letter *p* in *pig.*" Each consonant in Table 1.1 is designated by specifying voicing, place of articulation, and manner of articulation. Thus, *b* is a **voiced bilabial stop,** and *s* is a **voiceless alveolar fricative.** Since in English all nasal consonants are voiced stops, we describe *ng* simply as a **velar nasal.** Similarly, *r* may be described as a **post-alveolar approximant** without reference to voicing.

Table 1.1
Table of Examples of English Consonants Showing Place and Manner of Articulation

	Bilabial	Labio-dental	Dental	Alveolar	Post-alveolar	Palatal	Velar	Glottal
stop								
vcless								bottle[b]
asp[a]	Paul			tall			keg	
nonasp	spill			stall			skull	
voiced	ball			doll			gale	
nasal stop	mat			not			ring	
fricative								
vcless		fan	thigh	sink	shy			heaven
voiced		van	thy	zinc	vision			
approx-imant	win[c]			leap[d]	reap	year	win[c]	
affricate								
vcless						chug		
voiced						jug		
flap				ladder				

[a] *asp* and *nonasp* are *aspirated* and *nonaspirated*. Voiced stops are not usually aspirated in English.

[b] The glottal stop occurs in this word in some dialects of English. It is voiceless.

[c] The *w* sound in English has two approximate obstructions, one at the velum and one at the lips. You should feel your lips pucker when you articulate the first sound in *win*.

[d] The *l* sound is often articulated by raising the tongue tip to the alveolar ridge, while the **sides** of the tongue rest on the back teeth. For this reason it is called a **lateral approximant** by some linguists.

Note: The empty cells are either impossible sounds, such as a bilabial flap, or sounds that do not occur in English, such as a velar fricative (the final consonant in Bach enunciated in proper German).

1.2.2.2 Vowels

Speech sounds in which the airstream is unobstructed are called **vowels.** Vowels are more prolonged than consonants. You can sing vowels. Most of what you hear when you listen to opera are vowels, which is what makes the words of an opera difficult to understand. As a group, vowels are more homogeneous than consonants, which vary widely from stops with short durations of 60 milliseconds, to nasals, with which you can hum a tune, to approximants which themselves have vowel-like qualities.

How many vowel sounds are there in English? By now, I hope, you are suspicious of going by the alphabet and answering "five, *a, e, i, o, u.*" That this is wrong is easily seen by considering the following words:

beet, bit, bate, bet, bat, but, baht, bot, bought, boat, put, boot, abbot, bite, bout, boil

Each of the underlined vowels is pronounced differently from every other, and the differences distinguish the word. They are, therefore, each of them, separate vowel sounds of English.[4]

Vowels are sounds that resonate in the vocal tract, much like the air in a woodwind. In an instrument such as a clarinet, different tones are produced by using keys to shorten or lengthen the tube. In speaking, the tongue and lips perform a similar function, producing different vowels by altering the shape of the vocal tract so that the vibrating air produces sounds in which different frequencies are emphasized.

From an articulatory point of view, vowels are classified by the position of the body of the tongue and the lips during the production of the vowel. The tongue varies over two dimensions: **high-low**—toward or away from the roof of the mouth; and **front-back**—toward or away from the teeth. The lips may be **rounded** (or puckered) or **unrounded.**

The contrast in tongue height from **high** to **low** may be sensed when the following words are pronounced: *beet, bit, bate, bet,* and *bat.* Speak these five words and you will feel your jaw lower to accommodate the descending tongue body as it proceeds through **high, high-mid, mid, low-mid,** and **low** positions.

These vowels are all front vowels—the tongue is more or less right behind the teeth during their pronunciation. A five-some of back vowels, ranging in height from high to low, produces a similar, though somewhat less perceptible effect. Compare the difference in the openness of your mouth as you go through the list: *boot, put, boat, bought, pot.* (For some readers, the last two may be pronounced similarly. That is normal for your dialect.)

---

4. If *put* were pronounced with an initial *b,* it would not be mistaken for any other word in the list, so its vowel is distinct from all the other vowels. Also, *baht* is a unit of Thai currency. Its vowel is the same as in *balm.* The word *bot* refers to the larva of the botfly. Its vowel is the same as in *pot.* The vowel in the second syllable of *abbot* only occurs in unstressed syllables, and as such, it, too, is distinct from all the others. Some speakers may pronounce certain pairs of these vowels alike—for example, those in *bot* and *bought,* which are both pronounced as an "average" of the two. As long as other speakers distinguish between these two vowels, however, there is a need to represent them distinctly when analyzing English.

To experience the contrast in front-back, say *bat, but, bought.* It's subtle but you should feel your tongue body move from **front** to **central** to **back,** away from the teeth and toward the throat, as you proceed from *bat* to *bought.* The lowered position of your jaw as you say these words shows that the three are also low vowels.

The front-back contrast may also be felt with the high vowels in *beet* and *boot.* Try saying these in alternation: *beet, boot, beet, boot, beet, boot.* Your tongue shifts forward and back, but this effect may be masked by another effect, namely the rounding and unrounding of your lips. The vowel in *boot* (and *book, boat, bought*) is articulated with the lips rounded. For the vowel in *beet,* and all the front vowels, the lips are unrounded. You may see this by pronouncing these words as you look in a mirror.

The vowel in the second syllable of *abbot* is a "neutral" vowel, being neither high nor low, front nor back. It is the vowel that occurs in English in unstressed syllables. It also has a tendency to show up in casual, rapid speech.

The vowels in *bite, bout,* and *boil* have a slightly different character. During their articulation, the tongue moves from one position to another. From low-central to high-front in the case of *bite;* from low-back to high-back in *bout;* and from mid-back to high-front in *boil.* These three vowel sounds are **diphthongs.** They begin with one vowel and gradually change into another vowel. By prolonging the pronunciation of any of these diphthongs, you can feel the motion of your tongue, reflected by a change in jaw position. (It may help to look in a mirror as you do this.) The beginning of *bite* finds the jaw lowered for the low tongue position. As you proceed (slowly) to the end of the diphthong, the jaw will gradually close as the tongue rises toward the palate. Repeat this experiment with *bout* and not only will you feel the tongue and jaw rise together, but the lips will gradually round since non-low-back vowels in English are pronounced with the lips rounded. The above experiment may not work for some dialects of English such as many spoken in the southern United States, where diphthongs are different. Southern readers may have to imitate their northern brethren to achieve the tongue positions I just described.

The mid vowels in *bate* and *boat* are in actuality somewhat diphthongized, though not nearly to the degree as the vowels described overtly as diphthongs. The articulation of the a sound in *bate* begins with the tongue in a mid-front position, but ends with the tongue in a high-front position. The *oa* in *boat* does the same, going from mid-back to high-back. This is an effect you can feel if you are adept at prolonging vowels and sensing the location of your tongue. There is much variation among speakers of English as to the degree of diphthongization of these vowels. Vowels may also be nasalized—articulated with air passing out both the mouth and nose—as noted previously in our discussion of the function of the velum in Section 1.2.1.

Vowels can be charted on the basis of tongue height, tongue backness, and lip rounding. Table 1.2 shows the vowels of (some versions of) English including the dialect spoken by most national news broadcasters. The underlined letter(s) represents the vowel of interest. The rounded vowels of *boot, put, boat,* and *bought* are in boldface italics.

Vowels may be described by reference to Table 1.2. Thus the vowels in *beet* are **high-front vowels**; the vowel in *pot* a **low-back vowel**; the vowel in *bet* a **mid-low front vowel**; the vowel in *put* a **mid-high back vowel**; the vowel in *but* a **mid-low central vowel**. In English, rounding need not be mentioned specifically since it is predictable. In languages such as French or German, where front vowels may be both rounded and unrounded, it would be necessary to specify rounding in order to avoid confusion.

1.2.3 Phonetic Alphabets

For the purposes of scientific rigor, the English alphabet is inadequate. We are not the first to make this observation, though it has perhaps been never so well expressed as by George Bernard Shaw:

Table 1.2
Examples of English Vowels Indicating Place of Articulation

	Tongue Front	Tongue Central	Tongue Back
Tongue High	b**ee**t (end of diphthongs in *bite* and *boil*)		b**oo**t (end of diphthong in *bout*)
Tongue High-mid	b**i**t		p**u**t
Tongue Mid	b**a**te	abb**o**t	b**oa**t (start of diphthong in *boil*)
Tongue Low-mid	b**e**t	b**u**t	b**ou**ght
Tongue Low	b**a**t	b**a**lm (start of diphthongs in *bite* and *bout*)	p**o**t

Notes: Tongue position for the diphthong in *bite* begins in the Tongue-Low/Tongue-Central cell and finishes in the Tongue-High/Tongue-Front cell; tongue position for the diphthong in *boil* begins in the Tongue-Mid/Tongue-Back cell and finishes in the Tongue-High/Tongue-Front cell; tongue position for the diphthong in *bout* begins in the Tongue-Low/Tongue-Central cell and finishes in the Tongue-High/Tongue-Back cell. All these tongue positions vary, sometimes widely, with different dialects of English. This table reflects, more or less, the dialects spoken by national television news reporters. Any of these vowels may be spoken in a nasalized manner, but that is not indicated in this table.

It was as a reading and writing animal that Man achieved his human eminence above those who are called beasts. Well, it is I and my like who have to do the writing. I have done it professionally for the last sixty years as well as it can be done with a hopelessly inadequate alphabet devised centuries before the English language existed to record another and very different language. Even this alphabet is reduced to absurdity by a foolish orthography based on the notion that the business of spelling is to represent the origin and history of a word instead of its sound and meaning. Thus an intelligent child who is bidden to spell debt, and very properly spells it d-e-t, is caned for not spelling it with a b because Julius Caesar spelt the Latin word for it with a b.

What exactly is it that I, and perhaps you too by now, find inadequate about the alphabet? There are several answers to this question:

1. Different letters and letter combinations may represent the same sound. Consider this bit of nonsense:

    He grieves that Caesar sees the people seize the beans
    Where silly amoebas hide the keys to the machines

    The boldfaced letters, **e, ie, ae, ee, eo, ei, ea, y, oe, ey,** and **i** all represent the same high-front "e" vowel. Similar situations occur with *ph, gh, ff,* and *f,* all of which represent the "f" sound in *photo, rough, huff,* and *fast;* and *j, dg,* and *g* which all represent the same sound in *jet, edge,* and *siege.*

2. The same letter may represent different sounds.

    My father wanted many a village dame badly.

    In each case the letter *a* represents a different vowel sound. In other examples *c* has a "k" sound in *call* but an "s" sound in *city.* The letter *g* is pronounced differently in *guest* than in *gist.* Likewise for *h* in *huge* and *hug* and *s* in *sure* and *sewer.* Speak these aloud to hear the difference.

3. Silent letters:

    i*s*land bom*b*   throu*gh* *g*nome de*b*t   s*w*ord   *p*sychic *k*nit
    bit*e*   *w*hole resi*g*n   *gh*ost   *m*nemonic autum*n* a*l*ms

    And this is true: A student enrolled under the name of Lynn VanScoyoc pronounced her last name "vanskoy," assuring people that the *oc* was silent.

4. Some sounds in some words are not represented by any letter. The word *use* actually begins with the palatal approximant y-sound. Similarly, there is an unrepresented y in *acute, feudal,* and in some people's pronunciation of *tune.*

5. Finally, as we've seen, a single letter may represent two sounds: *x* is the chief malefactor here. It's pronounced *ks* in *sex* but *gz* in *exam.* /

Whether animals, celestial bodies, or speech sounds are being studied, science generally prefers to give unique names to unique objects within the purview of study. Astronomers are careful to name different stars differently, and they do not use the same name twice. A violation of this one-name/one-object principle may be a sign of confusion. The ancients thought that they knew of two heavenly bodies, which they called *Phosphorus* (the morning star) and *Hesperus* (the evening star). Later astronomers realized these two objects were one and the same, namely the planet Venus. Similarly, in biology scientists go to great extremes in their use of Latin and Greek to ensure that *flora* and *fauna* (themselves Latin terms) are named uniquely.

The alphabet violates this one-symbol-for-one-object principle in the various ways enumerated in the above list. What is needed is a special alphabet in which each sound is represented by one and only one symbol. Such an alphabet is called a **phonetic alphabet.**

A phonetic alphabet is needed by specialists concerned with speech so that they have an unambiguous way of discussing the objects of interest, namely speech sounds. We've gotten along without a phonetic alphabet so far by using locutions such as "the first sound in *pig* is an aspirated voiceless bilabial stop," but we need a more concise means.

Before introducing a phonetic alphabet, some words need to be said in defense of our own alphabet and orthography. Perhaps George Bernard Shaw was right about the spelling of *debt,* but, on the other hand, the presence of the *b* serves to remind us of the cognate words *debit* and *debenture* where the *b* is actually pronounced. Likewise for the silent *b* in bomb which is overt in *bombard, bombardier,* and *bombastic.*

Many other so-called irregularities in spelling actually reflect deeper regularities in the language. The silent *g* in *sign* becomes manifest in the semantically related words *signature* and *signal;* the silent *k* in *knowledge* appears fullblown in the related word *acknowledge,* as does the silent *n* in *autumn* which is pronounced in *autumnal.*

Different spellings for the same sound help us to disambiguate possibly confusing sentences such as *the book was red* versus *the book was read.* Lewis Carroll, the author of *Alice's Adventures in Wonderland,* often used English spelling as a vehicle for humor, as in this passage:

"And how many hours a day did you do lessons?" said Alice.

"Ten hours the first day," said the Mock Turtle, "nine the next, and so on."

"What a curious plan!" exclaimed Alice.

"That's the reason they're called lessons," the Gryphon remarked, "because they lessen from day to day."

A closer examination of English reveals further regularities. One might complain that the letter *a* is used for two different vowel sounds in *sane* and *sanity*. However, the use of the same letter a reveals that both words are related, as it does in pairs like *profane/profanity* and *humane/humanity*. Similar situations exist with *divine/divinity*, where the second i in each word is pronounced differently, or *serene/serenity* where the second e is the culprit.

Alternate pronunciations of related words can be a guide to correct spelling. Do you have trouble spelling *analog, melody, orchestra,* and *hygiene?* Do you find yourself wondering if they are spelled *anolog, melady, orchistra,* and *hygene?* You probably don't if you can pronounce the words *analogous, melodious, orchestral,* and *hygienic.* In these four the "true" values of the vowels are enunciated, and their spelling is obvious.

The International Phonetic Alphabet (IPA) [3] was devised by linguists to express unambiguously every speech sound of every language known in the world. This is the standard phonetic alphabet in use worldwide. Unfortunately, many of the symbols used for English speech sounds are unfamiliar to nonlinguists. Often, then, books such as this one introduce a phonetic alphabet of symbols more familiar to those likely to read the book. The principle of one-sound/one-symbol *must* be maintained, but the choice of symbol may be one of convenience, even to the point where two letters are used to represent a single symbol.

We have already done most of the work toward a phonetic alphabet. We have identified many important speech sounds of English, both consonants and vowels, as illustrated in Tables 1.1 and 1.2. We need only decide on symbols to represent them.

1.2.3.1 Phonetic Alphabet for English Consonants

Table 1.3 presents phonetic alphabet symbols for English consonants. Each pair of columns corresponds to a manner of articulation. Both oral and nasal stops, however, have been listed under **Stops**.

Several of these symbols require explanation. The **ʔ** is a glottal stop, the catch-in-the-throat sound that occurs in *uh oh*. It is also an alternate pronunciation, for some people, for the *tt* in words like *bottle* and *little*. Likewise the flap D is a common pronunciation for the *t, d, tt,* and *dd* that occur in words like *atom, Adam, letter,* and *ladder.* The "exponent" *h* denotes an aspirated consonant.

Table 1.3
A Phonetic Alphabet Illustrated by Example for the Pronunciation of Some English Consonants

Stops		Fricatives		Approximants		Affricates		Flap	
Symbol	Exemplar	Symbol	Exemplar	Symbol	Exemplar	Symbol	Exemplar	Symbol	Exemplar
p^h	Paul	f	fan	l	leap	ch	chunk	D	ladder
p	spill	v	van	r	reap	j	junk		
b	ball	th	thigh	y	year				
t^h	toll	dh	though	w	win				
t	stole	s	sink						
d	dale	z	zinc						
k^h	kill	sh	shy						
k	skill	zh	vision						
g	gold	h	heaven						
m	mock								
n	not								
ng	ring								
?	bottle								

Note: The digraphs—symbols with two letters—are to be considered as single, indivisible symbols.

The *dh* for the voiced dental fricative is necessary because English uses *th* for both the voiced and voiceless versions. Once upon a time English had a special letters for each of these sounds. If you ever have the opportunity to peruse ancient English texts, you may see them. The *zh* for the voiced palatal fricative is a compromise. The *z* tells you that the sound is voiced, and the *h* is supposed to remind you of the voiceless counterpart spelled *sh*. This sound is spelled with *z* as in *azure,* *s* as in *leisure,* *ti* as in *equation,* *g* as in *centrifuge,* *j* as in *rajah,* and *x* as in *luxurious.* (Note that the *h* of *dh* and *zh* has nothing to do with aspiration.)

1.2.3.2 Phonetic Alphabet for English Vowels

The vowels present a more complex situation. Table 1.2 has sixteen vowels, three of which are diphthongs, and ignores vowels that are nasalized. However, the English alphabet is particularly impoverished when it comes to vowel letters. That is why in spelling so many vowels are spelled with digraphs, such as *ai, au, ea,* and *ei.*

Ideally, we would use the IPA symbols. They were designed by an international committee of the world's best phoneticians over a period of decades. Unfortunately, too many of the vowel symbols they use are unfamiliar to the nonlinguist. Since this is not a book on linguistics per se we have the luxury of using symbols more in line with our experience. Two factors guided the creation of the phonetic alphabet used in this text. First, I wanted the diphthongs of *bite, bout, boil* to be represented by two letters, and the other vowels by a single letter, even though the two letters are to be regarded as a single symbol. Second, I wanted a high degree of familiarity. To this end, I have borrowed many of the phonetic symbols from dictionaries. Most dictionaries have their own phonetic alphabet to indicate the pronunciation of words, and readers are likely to be familiar with one. A key to the phonetic symbols occurs on every page in the dictionary where their values are indicated by exemplars. Thus, ā *pay* would describe the mid-front vowel. Table 1.4 presents a phonetic alphabet for English vowels.

The vowels with the macron, ā, ē, ō, ū have traditionally been called long vowels. Most American students are already familiar with their phonetic value. Unfortunately, the term *long* is also applied to the diphthong *ai,* which is called a "long i" and written as an *i* with a macron over it. The usage is inconsistent—the diphthong *ou* isn't considered long, although it is similar to *ai.* We will use the term *long* to refer to vowel duration, as discussed in Section 1.2.4. The other symbols correspond more or less to traditional dictionary phonetics and should be fairly recognizable.

1.2.3.3 The Phonetic Alphabet in Terms of Phonetic Properties

Tables 1.5 and 1.6 are consonant and vowel tables with phonetic alphabet symbols in place of words. From these tables you can read off the phonetic

Table 1.4
A Phonetic Alphabet Illustrated by Example for the Pronunciation of Some English Vowels

	Front Vowels		Central Vowels		Back Vowels		Diphthongs	
	Symbol	Exemplar	Symbol	Exemplar	Symbol	Exemplar	Symbol	Exemplar
High	ē	b*ee*t			ū	cr*u*de	ai	b*i*te
	i	b*i*t			u	p*u*t	ou	*ou*t
Mid	ā	bl*a*de	ə	it*e*m	ō	n*o*te	oi	b*oi*l
	e	b*e*t	ŭ	b*u*t	ô	s*aw*		
Low	ă	b*a*t	a	b*a*lm	o	p*o*t		
Note: Nasalized vowels may be denoted by a tilde exponent, for example i˜.								

description of any symbol. For consonants, the column determines the place of articulation while the row determines the manner of articulation, including voicing and aspiration. For vowels, the row and column indicate tongue height and backness. The roundedness of English non-low back vowels is denoted by bold italics; diphthongs are indicated by noting the points of articulation where they start and end.

1.2.3.4   Other Phonetic Alphabets in Common Use

This book emphasizes the **concept** of the phonetic alphabet—one sound, one symbol. Much of the literature in speech makes use of different symbols. Some use the IPA symbols already referred to; others use symbols from a phonetic alphabet called the **Arpabet,** which was devised for research projects funded by the Advanced Research Projects Administration (ARPA) of the Department of Defense. (More on ARPA in Chapter 3.) Tables 1.7 and 1.8 present the equivalent of this book's phonetic alphabet to the IPA, Arpabet, and modified IPA symbols (used by many American linguists). An excellent reference for all IPA symbols is found in [4].

1.2.3.5   Using the Phonetic Alphabet

In rendering English phonetically, we lose some fluency, both in reading and writing. Nevertheless, to be scientific about speech processing, both investigators and practitioners must acquire some skill in this area. The following is an example of what a phonetic transcription would look like using the phonetic alphabet of this book and using (more or less) the pronunciations of national news broadcasters.

Table 1.5
English Consonant Symbols Using the Phonetic Alphabet of This Book and
Showing Place and Manner of Articulation

	Bilabial	Labio-dental	Dental	Alveolar	Post-alveolar	Palatal	Velar	Glottal
stop								
vcless								ʔ[b]
asp[a]	pʰ			tʰ			kʰ	
nonasp	p			t			k	
voiced	b			d			g	
nasal stop	m			n			ng	
fricative								
vcless		f	th	s	sh			h
voiced		v	dh	z	zh			
approx-imant	w[c]			l[d]	r	y	w[c]	
affricate								
vcless						ch		
voiced						j		
flap			D					

[a]*asp* and *nonasp* are *aspirated* and *nonaspirated*. Voiced stops are not usually aspirated in English.

[b]The glottal stop occurs in some dialects of English. It is voiceless.

[c]The *w*-sound in English has two approximate obstructions, one at the velum and one at the lips. You should feel your lips pucker and the back of your tongue rise toward the velum when as you articulate the first sound in *win*.

[d]The *l*-sound is often articulated by raising the tongue tip to the alveolar ridge, while the sides of the tongue rest on the back teeth. For this reason it is called a **lateral approximant** by some linguists.

Now is the time for all good men and women to come to the aid of their country.
[nou iz dhə taim fôr ôl gud men ănd wimən tə kŭm tə dhə ād əv dher kŭntrē]

Phonetic transcriptions are placed in square brackets [ ] so that they are not confused with ordinary spelling. Also, there is no capitalization or punctuation. It's one symbol, one sound, no exceptions. There are several possible

Table 1.6
English Vowel Symbols Using the Phonetic Alphabet of This Book and
Showing Place of Articulation

	Tongue Front	Tongue Central	Tongue Back
Tongue High	ē (ai & oi end)		*ū* (ou end)
Tongue High-mid	i		*u*
Tongue Mid	ā	ə	*ō* (oi start)
Tongue Low-mid	e	ŭ	*ô*
Tongue Low	ă	a (ai & ou start)	o

Notes: Dipthongs *ai* of *bite, ou* of *bout,* and *oi* of *boil* are indicated by showing their tongue-start and tongue-end positions; boldfaced italicized vowels are rounded.

transcriptions of a given sentence, depending on how smoothly it is pronounced and how carefully the speaker is talking. For example, in a careful pronunciation the word *to* might be more faithfully rendered with its full vowel value, [tū]. In smooth flowing informal speech the ū is usually pronounced as the schwa ə. Similarly, *of* may be transcribed as [ŭv], [av], [ov], or [əv]. There will also be individual differences for people brought up in different parts of the United States or English-speaking world. Finally, for the sake of keeping the discussion from being overwhelmed by notation, I am overlooking the fact that the vowels preceding the nasal consonants, such as the [ai] in [taim] (time) are generally nasalized and could be so indicated by a tilde (~) superscript. Likewise, I have not distinguished between aspirated and unaspirated voiceless stops. In a narrower phonetic transcription, *time* would be [tʰai~m].

With these caveats in mind, let's look at another phonetic transcription.

The time has come	[dhə taim hăz kŭm]
The walrus said	[dhə wôlrəs sed]
To talk of many things	[tū tôk ŭv menē thingz]
Of shoes and ships and ceiling wax	[ŭv shūz ănd ships ănd sēling wăks]
Of cabbages and kings	[ŭv kăbəjəz ănd kingz]
And why the sea is boiling hot	[ănd wai dhə sē iz boiling hot]
And whether pigs have wings	[ănd wedhər pigz hăv wingz]

Table 1.7
English Consonant Symbols Using Various Phonetic Alphabets

This book	IPA	Arpabet	Modified IPA*	Description	Example
p	p	p	p	voiceless bilabial stop	s<u>p</u>ill
p^h	p^h	none	p^h	voiceless aspirated bilabial stop	<u>p</u>ill
b	b	b	b	voice bilabial stop	<u>B</u>ill
t	t	t	t	voiceless alveolar stop	s<u>t</u>ill
t^h	t^h	none	t^h	voiceless aspirated alveolar stop	<u>t</u>ill
d	d	d	d	voiced alveolar stop	<u>d</u>ill
k	k	k	k	voiceless velar stop	s<u>k</u>ill
k^h	k^h	none	k^h	aspirated voiceless velar stop	<u>k</u>ill
g	g	g	g	voiced velar stop	<u>g</u>ill
m	m	m	m	bilabial nasal stop	<u>m</u>ill
n	n	n	n	alveolar nasal stop	<u>n</u>il
ng	ŋ	ng	ŋ	velar nasal stop	si<u>ng</u>
ʔ	ʔ	q	ʔ	glottal stop	bo<u>tt</u>le**
f	f	f	f	voiceless labio-dental fricative	<u>f</u>ine
v	v	v	v	voiced labio-dental fricative	<u>v</u>ine
th	θ	th	θ	voiceless interdental fricative	<u>th</u>ick
dh	ð	dh	ð	voiced interdental fricative	o<u>th</u>er
s	s	s	s	voiceless alveolar fricative	<u>s</u>ip
z	z	z	z	voiced alveolar fricative	<u>z</u>ip
sh	ʃ	sh	š	voiceless post-alveolar fricative	fi<u>sh</u>
zh	d3	zh	ž	voiced post-alveolar fricative	a<u>z</u>ure
h	h	h	h	glottal fricative	<u>h</u>orse
l	l	l	l	(lateral) alveolar approximant	<u>l</u>eap
r	ɹ	r	r	(central) post-alveolar approximant	<u>r</u>eap
y	j	y	j	palatal approximant	<u>y</u>ear
w	w	w	w	labio-velar approximant	<u>w</u>olf
ch	tʃ	ch	č	voiceless palatal affricate	<u>ch</u>eer
j	d3	jh	ǰ	voiced palatal affricate	<u>j</u>eer
D	ɾ	dx	D	alveolar flap	la<u>dd</u>er

*The modified IPA is used, with some individual variation, by American linguists, for example [5].

**The glottal stop occurs in this word in some dialects of English.

Table 1.8
English Vowel Symbols Using Various Phonetic Alphabets

This book	IPA	Arpabet	Modified IPA*	Description	Example
ē	i	iy	i	high front vowel	m<u>e</u>te
i	ɪ	ih	ɪ	high-mid front vowel	b<u>i</u>t
ā	e	ey	e	mid front vowel	m<u>a</u>te
e	ɛ	eh	ɛ	low-mid front vowel	b<u>e</u>t
ă	æ	ae	æ	low front vowel	b<u>a</u>t
ū	u	uw	u	high back vowel (rounded)	l<u>u</u>te
u	ʊ	uh	ʊ	high-mid back vowel (rounded)	p<u>u</u>t
ō	o	ow	o	mid back vowel (rounded)	m<u>o</u>te
ô	ɔ	ao	ɔ	low-mid back vowel (rounded)	b<u>ough</u>t
o	ɒ	aa**	a**	low back vowel	p<u>o</u>t
ə	ə	ax	ə	mid central vowel (unstressed)	abb<u>o</u>t
ŭ	ɜ	ah	ʌ	low-mid central vowel	b<u>u</u>t
a	ɐ	aa**	a**	low central vowel	f<u>a</u>ther
ai	ɑɪ	ay	aɪ	low-back to high-front diphthong	b<u>i</u>te
ou	ɑʊ	aw	aʊ	low-back to high-back diphthong	b<u>ou</u>t
oi	ɔɪ	oy	ɔɪ	mid-back to high-front diphthong	b<u>oi</u>l

*The modified IPA is used, with some individual variation, by American linguists, for example [5].

**These alphabets symbolize both the initial vowels of *pot* and *father* identically. In many dialects of English these vowels are pronounced the same, but there are some dialects in which they are pronounced differently.

Try rendering the following phonetic transcription into properly spelled English:

[dhə ôthər ŭv dhis buk iz robərt dē rodmən]

If you have the hang of it you will be reminded of this book's author.

Now try this one: [ə măn ə plăn ə kənăl pănəma]. You'll know you have it right if your answer is a palindrome, a saying that reads the same backward and forward in ordinary writing, ignoring spaces and punctuation.

Without looking at the answer in the footnote, try to transcribe *the quick brown fox jumped over the lazy dogs* using our phonetic alphabet.[5]

---

5. dhə kwik broun foks jŭmpt ōvər dhə lāzē dogz

1.2.4 Prosody and Suprasegmentals

All speech, even the commonest speech, has something of song in it.

Thomas Carlyle

Imagine singing the first line of the song "Happy Birthday": "Happy birthday to you." The printed version doesn't really convey the overall sound very well, even in phonetic transcription: [hăpē bərthdā tū yū]. It gives you the individual consonants and vowels, or the phonetic segments of each word, but it doesn't convey the fact that *birth-* has a higher pitch than *happy* or *-day*, that *to* has an even higher pitch, and that *you* is drawn out at the end of the line. It also fails to indicate that *birth* is the most stressed syllable in the line. A better rendition might look like this:

> to
>
> youuuuuuu
>
> BIRTH
>
> Happy   day

In singing it is obvious that **pitch, length,** and **stress** are present. These are **prosodic** features. They are also called **suprasegmental** features in the sense that they are imposed over the segments. These qualities are present in speech as well as in singing.

Nobody speaks without varying pitch unless they deliberately speak in a monotone. Even within a single word, spoken normally, the pitch may vary. Most people say *teacher* with a higher pitch on *teach-* than on *-er.* The overall pitch contour of a sentence, called **intonation,** not only keeps speech from being stupefying but may actually alter meaning. The sentence, "She's leaving," spoken with falling intonation is a statement, whereas with a rising intonation it's a question.

All sentences in normally spoken English have an intonation pattern determined by the words, their position in the sentence, and their meaning. Often, in simple sentences, the intonation peaks at the verb, as in

> love
>
> I   you.

In other cases it's more subtle. If the sentence, "Koala bears eat eucalyptus leaves," is in response to a question of what kind of animal eats eucalyptus leaves, the intonation pattern peaks over *koala bears:*

    ala be

Ko   ars eat eucalyptus leaves

If it is in response to the question of what koala bears eat, the intonation pattern peaks over *eucalyptus leaves:*

        lyp

Koala bears eat euca   tus leaves.

For English, in general, words conveying the most significant information from the point of view of the speaker receive highest intonation, other things being equal.

In most polysyllabic words of English, some syllables receive more emphasis than others. In *fortitude* the first syllable *for* is emphasized most, the second syllable *ti* is emphasized least, and the third syllable *tude* has medium emphasis. (You can determine this by mentally shouting the word. You will discover you put the most energy into the first syllable.) We say that *for* has **primary stress** and *tude* has **secondary stress.**

Stress plays an important role in English. Consider the pair of words *melody* and *melodious.* Because the syllable containing *o* in melody is unstressed, the *o* is pronounced as schwa [ə]. In general, unstressed English vowels are pronounced as schwa, if they're not omitted altogether. Because of its internal structure, the word *melodious* receives primary stress on the syllable containing the *o,* so that vowel is pronounced with its full value.

Stress may even make a difference between two words containing exactly the same segments. Consider the verb *convert* and the noun *convert.* Pronounce both out loud in the following sentences:

- Will she convert and become a pagan?
- He is a convert to paganism.

As a verb, *convert* receives primary stress on the second syllable, as a noun, on the first. There are many such pairs in English: *insult, pervert, reject,* etc.

Stress may be used for contrastive emphasis. Consider two sentences. Boldface indicates contrastive stress:

- **Beavers** build dams.
- Beavers build **dams.**

The first might be used in the sense of "**Beavers** build dams, not platypuses," the second in the sense of "Beavers build **dams,** not highways." Almost any word of a sentence may receive contrastive stress, which is imposed over the normal stress patterns of the sentence. Even parts of words may receive contrastive stress as in, "That's my employ**ee,** not my employ**er.**"

Stress is significant on the phrase level. A *hotdog* is a frankfurter, while a *hot* **dog** is an overheated pooch. The president lives in the *White House* which happens to be a *white house.* A *light housekeeper* is a skinny servant; a *lighthouse keeper* minds a warning beacon for ships.

Even on the sentence level the placement of primary stress may affect meaning. Consider the difference between these two sentences:

- I left **directions** for you to follow.
- I left directions for you to **follow.**

In the first, I wrote down some instructions for you to carry out regarding, say, the use of my computer. In the second I asked you to follow me somewhere.

The length of a segment may be increased by prolonging its articulation. In the case of vowels, fricatives, nasal stops, and approximants, this is accomplished by holding the sound for the desired duration. For stops and affricates, length is achieved by holding the closure for additional milliseconds. In English, length is not a distinctive feature. Prolonging a vowel as in *I'm sooooooo tired* doesn't change the meaning of *so,* although it may convey the additional sense of how terribly tired I may feel.

In Japanese, on the other hand, vowel length (duration) is distinctive, so a word in which a vowel is pronounced in a clipped manner could be a different word than if the same vowel were prolonged. I once taught for a semester in Japan and learned about vowel length in a rather amusing way. I called role every day, and when I called Ms. Tsuji the class would titter and Ms. Tsuji would blush. After a few days of this I asked another professor what was going on. He knew right away. The word *tsuuji* with the *u* prolonged meant "evacuation of the bowels." With a short vowel it was a person's surname. Having been living in the South for some years, my vowels, even when pronouncing Japanese, were drawled, and my pronunciation of Ms. Tsuji's surname was missing the mark. Ironically, my first correct pronunciation evoked peels of laughter from the class as they instantly knew that I knew. Even poor Ms. Tsuji was smiling.

It happens in English that certain vowels are held slightly longer than others. The difference is difficult to sense unless you're a trained phonetician, but machine measurements show it to be true. Thus $\bar{e}, \bar{a}, \bar{u}, \bar{o}, \hat{o}, a, ai, oi, ou$ are held slightly longer than *i, e, ă, u, o, ŭ* and of course *ə*. This doesn't mean you can't stretch out the vowel [u] in *good* by saying *goooooooood boy* making the [u] longer than the [oi] of *boy*. The length differences occur in natural, free flowing speech.

Linguists call the long (i.e., lengthy) vowels **tense** and the short ones **lax**. This has little to do with tension and relaxation, however. The tense vowels are capable of occurring at the ends of one-syllable words, such as *see, say, Sue, sow, saw, pa, buy, boy, bough* which are, phonetically [sē, sā, sū, sō, sô, pa, bai, boi, bou]. One-syllable words cannot end with a lax vowel in English. Try saying [bi] or [be]; it feels unnatural. A deeper discussion of these matters may be found in [2].

In summary, the prosodic properties of speech are pitch, intonation, stress, and length. In some cases, their use changes the meaning of a word, phrase, or sentence. In other cases, it conveys emotional or other connotations. Both computer speech recognition and synthesis must take these suprasegmental properties into account to be effective, as we shall see in Chapters 3 and 4.

1.2.5 Syllables

The syllable is arguably the smallest utterable unit of speech. Every utterance contains at least one syllable. In English, a syllable may consist of a vowel alone; a vowel preceded by one, two, or three consonants; a vowel followed by one, two, three, or four consonants; and combinations of these. The following words all contain one syllable:

- *Owe:* A vowel alone;
- *Me:* A vowel preceded by a single consonant;
- *Am:* A vowel followed by a single consonant;
- *Strew:* A vowel preceded by three consonants;
- *Inks:* A vowel followed by three consonants;
- *Strengths:* A vowel preceded by three and followed by four consonants.

Most syllables contain a vowel, called its **nucleus**. Any consonants preceding the vowel are the **onset** of the syllable. Any consonants after the vowel are the **coda** of the syllable. Syllables may lack an onset, a coda, or both.

Not all syllables need have a vowel as the nucleus. Consider words like *button, bottom, haggle,* and *mutter.* Many speakers pronounce them phonetically as [bŭtn], [botm], [hagl], and [mŭtr], without a vowel between the final two segments.[6] These words have two syllables so we must allow [n], [m], [l], and [r] to be syllable nuclei on their own. These four sounds play a dual role in English syllable structure, as they may function both as onsets and codas, as well as nuclei.

English has many restrictions on what sounds may occur together. Crossword puzzle addicts are aware of these restrictions on the spelling level. In working a crossword puzzle, if you know the first letter of a word is *r* you can be fairly certain the next letter is a vowel. If the last letter is *t* the next-to-last letter is unlikely to be *m*. These restrictions are confined to the syllable. For example, an [r] may be followed by an [l] if they occur in different syllables, as in *hairless.*

People are more aware of syllables than they are of segments—the individual consonants and vowels. If you ask a person how many sound segments in a word they will usually tell you how many letters there are in the spelling of the word. Even when you insist on making them count the number of individual **sounds** they often can't do it. However, most people can tell you precisely how many syllables there are in a word, once they grasp the concept of a syllable, which most do easily.

Awareness of syllables is borne out historically. Writing systems in which one symbol represented one syllable (called a **syllabary**) preceded alphabetic writing systems in which one symbol represented (by-and-large) one sound segment. The first alphabet, devised by the ancient Greeks, and the source of our own alphabet, was a modification of a Semitic syllabary that had been evolving for thousands of years.

Pitch, stress, and length are all properties of the syllable, rather than of individual segments. Intonation, which extends beyond the syllable to larger units such as words and phrases, is the aggregate effect of syllable pitch smoothed naturally by the speaker much as a singer makes a smooth transition between the individual notes of a song.

1.2.6 Dialects

A **dialect** of a language is the way a certain group of people, often distinguished geographically, renders that language. Dialects may vary in pronunciation as well as terminology and even grammar. Our concern is mainly with pronunciation.

---

6. Some speakers pronounce these words with a schwa in the last syllable. They would be transcribed as [bŭtən], [botəm], [hagəl], and [mŭtər].

Dialect must not be taken to mean some inferior or otherwise degraded form of a language. Everyone speaks some dialect of their native language. Even the Queen of England speaks a dialect of British English known as "received pronunciation." In America national broadcasters learn to speak a "broadcaster's dialect," which is no more or less correct than the English spoken by native southerners, midwesterners, or easterners. Variety in language is a fact of nature, as natural to language as variety in botany or geology. The dialect a person speaks is not a reflection on that person's intelligence or level of education, though it may provide clues to social standing, place of birth, and current location within the country.

We are concerned with dialects because when we decide how to transcribe a certain word phonetically, we are implicitly referring to some dialect. For example a Californian would transcribe the word *Cuba* as [kyūbə]. A Bostonian would transcribe it as [kyūbər]. A southerner enjoys apple [pa], a northerner apple [pai]; [yū sā təmātō ănd ai sā təmatō] is a line in an old song that spoofs the two different ways of pronouncing *tomato* found in two English dialects. Is one correct and the other wrong? Not at all; both are correct with respect to the particular dialect.

The consonants and vowels described in this book are for a group of dialects spoken in the United States. These are the dialects described in most dictionaries of American English, insofar as pronunciation is concerned. The broadcaster's dialect is among them. For speakers of other dialects, probably the majority of readers, some things are going to feel wrong. Diphthongs, especially, are problematic. Some dialects do not have certain diphthongs where they are described in the phonetic chart, and do have diphthongs where I haven't mentioned any. To all of you I apologize. However, my point is only to illustrate the concepts of phonetics and phonetic alphabets, not to aggrandize one particular dialect.

People also speak in different **styles.** In a sense, these are dialects too, but they're not based on region or social class. Rather they depend on the situation of the moment. All of us have a formal style that we might use when interviewing for a job or conversing with a senator, and we have an informal style that we use with family and close friends. Formal styles tend to avoid contractions—*can not* rather than *can't*—and do not drop vowels—[jenərəl] (*general*) rather than [jenrəl].

Many people have other styles that they use on different occasions (subconsciously). For example, Hispanic speakers of English use a different style when speaking with Anglos than with other Hispanics. People often speak differently to small children or to their pets. Some individuals even use a different style when talking to answering machines or computers, often a loud, carefully enunciated formal style of the kind one might use when conversing with a child.

Humans are generally not disturbed by different styles or dialects with which they are unfamiliar. Usually a moment of adjustment will allow an Iowan to understand a Georgian, and vice versa. Computers are a different story, and for those of us interested in computer voice technology, styles and dialects are among the more significant challenges. Much effort goes into making computers robust when processing different variations of speech. Moreover, anyone in speech processing who is unaware of, and doesn't understand the nature of these variations, is at a severe disadvantage.

1.2.7 Languages (Other Than English)

The world is becoming very internationalized, and this is no less true in computer speech technology. Natural language processing systems are increasingly being designed to be portable from language to language. Much effort is put into computer programs that translate and interpret between languages.

Different languages have different speech sounds. For example, the dental fricatives *th* and *dh* of English are not found in many of the world's languages. On the other hand, some languages have velar fricatives, the final, throat-clearing sound of Bach, which most dialects of English lack. (All English dialects once had this sound; it occurred in many words spelled today with *gh* such as *night*.) However, this lack of phonetic uniformity needn't deter us from dealing with non-English languages.

Despite the seemingly monumental differences between languages—most deeply felt when adults try to learn a foreign language—the similarities outweigh the differences. This fact was realized as early as the thirteenth century by Roger Bacon who wrote:

> He that understands grammar in one language understands it in another as far as the essential properties of Grammar are concerned. The fact that he can't speak, nor comprehend, another language is due to the diversity of words and their various forms, but these are *the accidental properties of grammar.*

All languages have their own inventory of speech sounds that are combined to form syllables and words. The words themselves are combined to form phrases and sentences, leveraging, in effect, the expressive capabilities of a relatively small number of speech sounds. The types of sounds are the same across languages—for the most part. All languages have stops, fricatives, and nasals. All languages use the lips, the palate, and the velum to articulate consonants. All languages contrast front with back and high with low vowels.

Even the grammatical rules of different languages show deep similarities, though the surface dissimilarities make them appear diverse. All languages have word classes such as *noun* and *verb*. All languages have rules for combining word classes into phrases and phrases into larger units, so that a commonality among languages is that there is no limit to the size of a sentence. This is clearly seen in the children's poem "The House That Jack Built." It begins with a single sentence that is repeated in the second verse. Then the whole thing recurs again, and so on. All languages have this recursive capability.

This is the house that Jack built.

This is the malt that lay in the house that Jack built.

This is the rat that ate the malt that lay in the house that Jack built.

This is the cat that killed the rat that ate the malt that lay in the house that Jack built.

This is the dog that worried the cat that chased the rat that ate the malt that lay in the house that Jack built.

What makes languages appear so different is a principle that they all share, which, ironically, argues for their universality. In all languages, the relationship between the sounds of a word and its meaning is arbitrary. There is no connection between the sounds of *dog* and the animal denoted by the word. Otherwise, why would the same creature be called *perro* in Spanish, *chien* in French, *hund* in German, *maa* in Thai, *inu* in Japanese? However, this makes words denoting the same things different in different languages. This, together with different sound inventories and different word orders, makes different languages appear to be more multifarious than they actually are.

With few exceptions, the consonants of the world's languages may be specified by voicing and place and manner of articulation. In Table 1.5, the occupied cells indicate the consonants in use for English and in many other languages as well. The empty cells are, for the most part, consonants that might have occurred in English, but through historical chance do not. They also represent sounds that may occur in other languages, for all languages may have their consonants catalogued in a chart of this kind.

For example, German has a labio-dental affricate. It is spelled *pf* in German and occurs in words like *pferd* meaning horse. Spanish has a palatal nasal, spelled *ñ* as in *señor.* Hungarian has both voiced and voiceless palatal stops.

Similarly, the vowels of the world's languages are aptly described in a chart similar to Table 1.6. Many languages familiar to us have front rounded vowels, absent in English. The *u* in French *lune* is a high, front rounded vowel, an [ē] articulated with the lips pursed. Other languages have diphthongs not utilized by English. On the other hand, English has vowels not found in neighboring languages. Spanish has a high-front vowel [ē] but lacks a mid high-front vowel [i], so that many native speakers of Spanish, when they learn English, pronounce *beet* and *bit* with the same vowel. Knowing such facts is crucial in designing speech processing systems that may be used by non-native speakers. (Analogous situations occur for English speakers learning Spanish and using Spanish computers.)

The words of all languages are made up of syllables. However languages differ in what comprises a syllable. As we've seen, English syllables may consist of a vowel alone, or a vowel surrounded by consonants and combinations of consonants. English is rich in syllable types. Other languages are more restrictive in allowable syllable types. In Hawaiian, a syllable may only be a vowel or a vowel preceded by a consonant as in *a-lo-ha, Ho-no-lu-lu, Ma-u-na Lo-a.* Japanese syllables are similarly restricted, except a syllable may end in a nasal consonant: *Ko-be, O-sa-ka, tem-pu-ra, Nip-pon.*

Most languages of the world (out of some 5,000) are **tone languages.** This comes as a surprise to most people. The Indo-European languages happen not to be tone languages (mostly), and those are the ones we are familiar with. Meanwhile, however, the thousands of Asian, African, and Native American languages, which comprise the bulk of the 5,000, are tone languages, and the language with the most speakers, Mandarin Chinese, is a tone language as well. In a tone language, the relative pitch on a syllable affects the meaning of the word. For example, in Thai the word that is segmentally *naa* means "rice paddy" when uttered with a mid tone, "younger sibling of mother" when uttered with a high tone, "face" when uttered with a falling tone, and "thick" when uttered with a rising tone.

The tones are relative within the utterance. Children with high voices and men with deep voices may still converse in their tone language, though the child's low tone may well be higher in pitch than the man's high tone.

English uses pitch in a different way, not to affect lexical meaning, but to convey such semantic information as inquiry or astonishment.[7] The use of pitch, whether in tone languages like Thai, or intonational languages like English, appears to occur universally in language.

---

7. There is a quibble here. The use of stress may affect lexical meaning in English, as in CONvert and conVERT. One of the acoustic correlates of stress is higher pitch, so in an indirect way one could argue that pitch affects lexical meaning in English. The quibble is important for us because we want a computer speech recognizer to distinguish such pairs, and pitch detection may play a role in accomplishing that.

Understanding that the differences among languages are incidental tells us that we can design speech processing hardware and software that will work for any language and therefore compete in the international marketplace. It also helps us design robust systems that are resistant to the speech of non-natives. Knowing that a French speaker of English is likely to substitute [z] for [dh] in words such as *the* because [dh] is not a sound in French helps us design a speech processor that can be successfully used by such speakers. As a by-product, such knowledge helps you learn to pronounce French or any other language. It's not that one articulates words in French by speaking English sounds with a "French accent." Rather, one learns the speech sounds of French outright, which is not a difficult task taken one sound at a time, and then uses that knowledge to pronounce the words of French.

1.3 Acoustic Phonetics

Articulatory phonetics is the study of the linguistic categories of sounds of human speech production, such as fricative, voicing, and bilabial. **Acoustic phonetics** is the other side of the coin. It is the study of the physical properties of the sound waves of speech that are pertinent to human language understanding.

Sound is composed of waves of pressure variations that oscillate from positive to negative relative to the surrounding medium, usually air. The number of air pressure oscillations each second determines the **pitch** of the sound, whose physical correlate is **frequency**. The size of the pressure variations determines the loudness of the sound, whose physical correlate is **intensity**. Most sounds, including speech sounds, are a complex mixture of such waves.

The term *pitch* is used to describe what humans hear. The term *frequency* is used to describe what scientists measure. Both refer directly to the rate of air pressure variations, high rates being perceived as a higher pitch and measured as a higher frequency, and vice versa. Pitch is often described as a musical tone, for example, A above middle C. Frequency is described in vibrations per second, or **hertz (Hz)**. The musical tone A above middle C is 440 vibrations per second, or 440 Hz.

Similarly, the term *loudness* is our perception of what scientists measure as *intensity*, itself proportional to the average amplitude of the sound pressure variations. Acoustic scientists prefer **relative intensity** as an indicator of loudness, which is measured in a base 10 logarithmic scale called **decibels (dB)**.

The decibel is defined as 20 times the base 10 logarithm of the ratio of the sound pressure level of the sound in question, to a reference sound pressure level that corresponds to the faintest sound perceptible to young, healthy ears.[8] If P represents the pressure of some sound S, and P_R the pressure of the reference sound, then the relative intensity of S in decibels is 20 log (P/P_R). The relative intensity of P_R itself is 0 dB, since the log of $P_R/P_R = \log 1 = 0$. A sound will have negative decibels if it is more faint than the reference sound, that is, below the level of normal human perception. It may be calculated that if one sound has twice the sound pressure level of another sound, its relative intensity will measure about 6 dB more than the other; three times the relative intensity corresponds to a difference of 10 dB, and ten times the relative intensity to a 20-dB difference. Conversely, a difference of 1 dB represents about a 12% difference in sound pressure level, which is at the threshold of detection by the human ear. (To calculate these relationships more easily, assume a $P_R = 1$. The intensities relative to each other will not be affected.)[9]

Based on the reference level P_R mentioned above, speech occurs in a range of 30 dB (whisper) to 80 dB (loud shouting), assuming the listener is within several feet of the sound. The perception of loudness is related to relative intensity in decibels—higher intensities are perceived as louder at a given frequency—but the relationship is not, in general, a linear one. This complex subject of loudness perception may be further explored in [6].

We can graph the waveform of a sound as pressure variation (amplitude) versus time. Figure 1.3 is a bit of waveform of the vowel [a]. Figure 1.4 is the same for [i].

The *y*-axis shows the total pressure variation, to which the loudness of the sound is related. For example, the *x*-axis measures time, a total of 0.02 seconds (20 ms) in Figures 1.3 and 1.4. The extreme peaks and valleys in air pressure—seen most clearly as the two negative minima in Figure 1.3—indicate the rate of vibration of the vocal cords. The time difference between the two minima is about 0.004 seconds, corresponding to 250 Hz, typical of a female speaker. There is too much vagueness in these representations for them to be usefully informative. The difference in the waveforms of [a] and [i] in Figures 1.3 and 1.4 does not appear to be related to their phonetic difference.

---

8. The reference level is somewhat arbitrary as even healthy young ears vary in their ability to perceive faint sound. The particular definition of decibels given here is called sound pressure level (SPL) dB, because it is related to the pressure wave. Another definition of decibels may be found in the literature based on the relative power of waveforms and is defined as 10 times the logarithm of a ratio of powers. The two definitions are essentially equivalent as power is proportional to the square of voltage, which is the unit used to measure SPL.

9. Here is one such calculation worked out. Suppose a sound P has twice the sound pressure as a sound Q ($P/Q = 2$). The relative intensity of P in decibels is $20(\log(P))$, and of Q is $20\log(Q)$. The difference in dB is $20(\log(P) - \log(Q))$. By the rules of logarithms this expression is the same as $20(\log(P/Q)) = 20\log(2) = 6.01$.

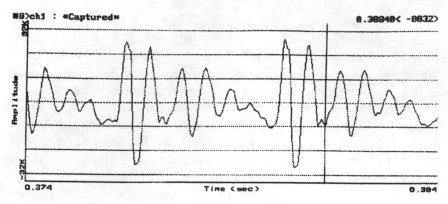

Figure 1.3 Waveform for a 10-ms segment of the vowel [a].

To determine the phonetic properties that allow listeners to distinguish one sound from another, we must decompose the complex wave form into the individual waves that make it up and examine their frequency and amplitude. This feat may be accomplished on a computer as a **spectrographic analysis.** Such an analysis reveals that each vowel has certain frequency bands with markedly high amplitudes or energy. For example, in a particular pronunciation of [ē], frequencies centered about 280 Hz, 1,960 Hz, and 2,660 Hz have much higher amplitudes than frequencies outside those ranges. On the other hand, [o] has maximum energy in frequency ranges of about 700 Hz, 1,120 Hz, and 2,240 Hz. We humans perceive these differences as differences in vowel quality.

The bands of high energy frequencies that occur in vowels are called **formants.** The lowest one is the **first formant (F1);** the next one is the **second formant (F2),** then comes the **third formant (F3)** and so on. Physically, the

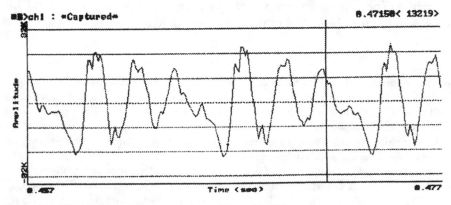

Figure 1.4 Waveform for a 20-ms segment of the vowel [i].

formants correspond to certain **harmonics** of the fundamental frequency produced by the vocal cords. (Harmonics are sounds whose frequencies are multiples of the fundamental frequency. For example, the formants of [o] given above are centered around the 5th, 8th, and 16th harmonics.) The fact that different vowels have different formants is the result of resonances in the vocal tract, whose shape varies as the tongue moves into position for the various vowels.

The formants cannot be seen in the waveforms of Figures 1.3 and 1.4, because they show the total effect of all the frequencies that compose the sound. A spectrographic analysis allows us to separate out the individual frequencies and to plot them and their amplitudes versus time. This results in a three dimensional display, which, while accurate, is difficult to comprehend. A better visual representation is the **spectrogram**. It is a plot of frequency on the y-axis, versus time on the x-axis. The third dimension, amplitude or energy, is represented on a gray scale by degree of darkness. The darker a frequency time point is, the higher the amplitude.

Figure 1.5 is a spectrogram of the vowels [ē] as in *beet*, [a] as in *balm*, and [ū] as in *boot*. In the spectrogram of [ē] the first four formants are visible as dark horizontal bands. The first formant is centered around 250–300 Hz, the second around 1,950 Hz, the third around 2,600 Hz, and the fourth around 3,100 Hz. The first four formants of [a] are also visible, but only the first and second formants of [ū] have enough energy to be displayed.

As you can see from Figure 1.5, formants represent a range of frequencies in which energies are highest. One can only assign a definite frequency to a formant by some kind of averaging or centering technique. Moreover, even within the articulation of a single vowel, the formants are in flux, mostly because the tongue is constantly moving as we speak, continually altering the shape of the vocal tract.

Vowel quality is determined mostly by the first and second formants. The third formant plays a small role, and higher formants appear not be significant. The first formant is inversely proportional to the height of the vowel, so it is lower for the high vowels [ē] and [ū] than the low vowel [a]. The second formant frequencies are higher for front vowels than back vowels of the same height. The degree of backness of a vowel of any height is more accurately reflected in the size of the difference between the first two formants: the smaller the difference the greater the backness. The **relationships** among the formants relate to vowel quality as much as the actual frequencies themselves. That's why men, women, and children can understand one another despite the fact that the absolute values of their formant frequencies vary.

These relationships are general tendencies that may not be fully revealed in a particular set of measurements. That is because there is much variability in the way we speak. Table 1.9 shows the values of the first three formants of

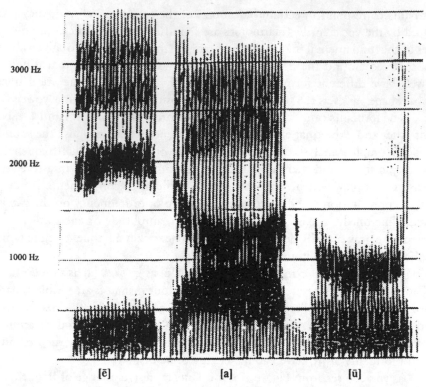

Figure 1.5 Spectrogram for the vowels [ē] as in *beet,* [a] as in *balm,* and [ū] as in *boot.*

twelve vowels as spoken by a large male speaking a dialect tinged with influence from Boston, Los Angeles, and Raleigh, North Carolina. These values must be taken as indicative only—the result of averaging values from a single session of speaking.

The data are more instructive when graphed. Figure 1.6 plots F2 on the vertical axis and F1 on the horizontal axis. The scale on the horizontal axis is

Table 1.9
Formant Frequencies of English Monophthongal Vowels for a Particular Speaker at a Single Session of Data Collection

	Frequency in Hertz											
F3	2620	2330	2300	2240	2200	2500	2050	2150	2250	2300	2200	2250
F2	1950	1720	1710	1590	1500	1300	1410	1100	920	1010	1050	850
F1	250	360	480	490	620	540	510	660	540	490	420	300
	ē	i	ā	e	ă	a	ŭ	o	ô	o	u	u

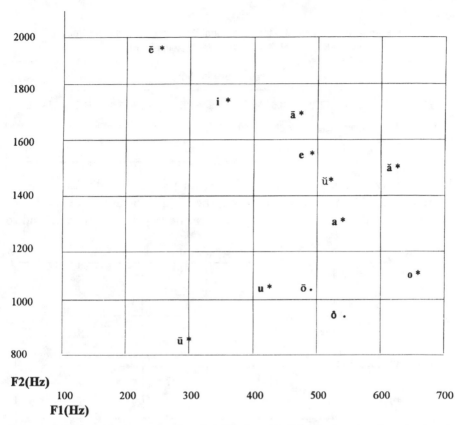

Figure 1.6 A plot of F2 versus F1 of twelve American English vowels, taken from the data in Table 1.9.

twice that of the vertical axis in order to spread the points for easier viewing in the formant space.

The more to the left a vowel is (small F1), the greater its tongue height; the more to the right it is (larger F1) the lesser its tongue height. Thus, the four leftmost vowels, [ē, ū, i, u] are articulated with more tongue height than vowels to their right, while the four rightmost vowels are articulated with less tongue height than vowels to their left. The topmost vowels tend to be front vowels while the bottom-most are back vowels, revealing the correlation between the second formant and tongue backness. A more striking relationship is apparent when we look at the difference between the first two formants plotted against the first formant. These are shown in Table 1.10 and Figure 1.7.

Smaller F2-F1 values indicate backness; larger values indicate frontness. Thus, all the vowels whose F2-F1 is below 700 are back vowels; those whose F2-F1 is above 1,000 are front vowels. In between are the vowels [ŭ], [a], and

Table 1.10
Formant Frequencies of English Monophthongal Vowels for a Particular Speaker at a
Single Session of Data Collection Showing F2-F1 and F1

	Frequency in Hertz											
2-F1	1700	1360	1280	1100	880	760	900	440	380	520	630	550
F1	250	360	480	490	620	540	510	660	540	490	420	300
	ē	i	ā	e	ă	a	ŭ	o	ô	o	u	u

[ă] (of *cut, calm,* and *cat*). The vowels [ŭ] and [a] are generally articulated as central vowels. The centralness of [ă] may be an idiosyncrasy of the speaker or of that particular utterance, or simply due to the fact that low vowels tend to

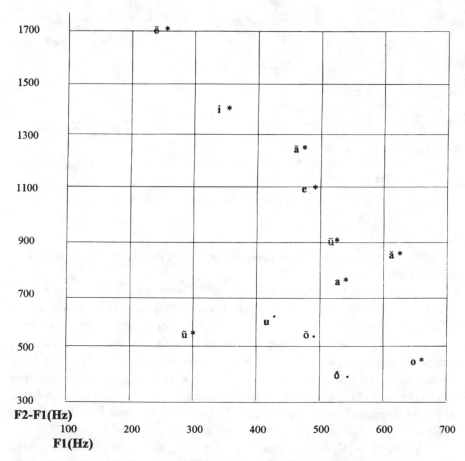

Figure 1.7 A plot of F2-F1 versus F1 of twelve American English vowels, taken from the data in Table 1.10.

be more central than high vowels because there is less room in the bottom of the mouth for the tongue to move back and forth than in the top.

The consonants adjacent to a vowel have a subtle effect on the vowel's formants. This effect is part of what we perceive in hearing a particular consonant. Consider the spectrogram of *bib* [bib] and *gig* [gig] in Figure 1.8.

Bilabial consonants such as [b] affect an adjacent vowel by lowering the normal locus of the second and third formants. This can be seen in the spectrogram of *bib*, where most of the second formant lies below the 2,000-Hz line. Compare this with the spectrogram of *gig* where the second formant of the [i] lies entirely above the 2,000-Hz. line. Typical of vowels adjacent to velar consonants such as [g] is the near common origin of the second and third formants, also evident in the spectrogram of *gig*.

Other spectrographic properties of consonantal features have been determined by phoneticians. Voiced sounds are characterized by dark vertical striations that correspond to the pulsing of the vocal cords. Stop consonants will produce a strip of vertical nothingness because nothing happens for some dozens of milliseconds. This will be followed by a burst of noise if the stop is voiceless, which has a mottled appearance in the higher frequencies, often in the range 3,000–6,000 Hz. A voiced stop, on the other hand, will terminate in the beginning of the formant structure for the following sound.

Nasals, as you might guess, have formant structure determined in part by the shape of the nasal cavity. These formants are weaker than those of oral vowels and show up faintly in spectrograms. The first, second, and third formants of nasal consonants occur at about 300, 2,600, and 3,300 Hz. Nasals also exhibit **antiformants,** which are frequencies of energy minima. These, of course, are not visible in a spectrogram but would appear in the spectrum, a kind of

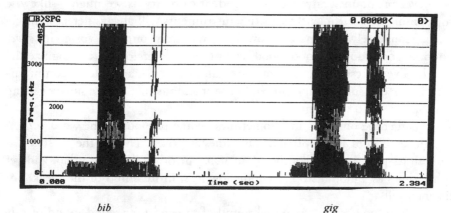

bib   gig

Figure 1.8 Spectrograms for the words *bib* and *gig*.

display discussed in Chapter 2. (Chapter 5 of [7] is a good source for the reader interested in more detail.)

Approximants have a vowel-like character. [r] and [l] may serve as the syllable peak as in the final syllables of *butter* and *bottle*. [w] and [y]—traditionally called "semi-vowels"—often function to connect one vowel to another as in such words as *Bowie* [bōwē] and *mayonnaise* [māyənāz]. Thus, it is not surprising that approximants have formant structures that resemble those of vowels.

A more complete discussion of the relationship between spectrograms and speech sounds may be found in [2] and [7]. We will return to spectrograms in Chapter 2 from a digital signal processing perspective.

1.4 Phonemics

If we speak the same word twice—"goodbye, goodbye"—the two utterances will be physically different. That can be easily shown with instruments. Despite this, we recognize it as the same word. Similarly, if we observe a person we know from the front or from the side, we recognize what we see as that person, though the two views are physically different. It seems as if we are able to go to the core of the matter, ignoring inessential differences, and focusing on distinctive characteristics. This ability is an integral part of human perceptual psychology.

Sometimes small differences are significant. The difference in pronunciation between *good* [gud] and *could* [kud] is quite small, smaller than that of the two pronunciations of *goodbye* such as "goooooodahbyyyyye" and "goo-by," yet the small difference is linguistically significant, whereas the large one isn't.

The ability to know when small differences are significant and large differences are not, is vital when it comes to speech. It can be verified that we never pronounce any two vowels identically, try as we might—likewise for consonants. Sufficiently sensitive instruments will always detect a difference. Conversely, no matter how long we draw out the vowel in *good,* gooooooooooooooooooooooood, even to a little melody, we recognize the word. Given these facts, it seems miraculous that we are able to communicate via speech sounds, and probably is miraculous that we are slowly getting computers to do the same.

Earlier we observed that sometimes people pronounce the word *little* as [litəl], sometimes as [liDəl] and sometimes as [liʔəl]. That is, the "t" sound is sometimes pronounced like the "t" in *stop,* but other times as a flap, and yet other times as a glottal stop. Yet somehow it is, in the mind of the speaker or listener, still a "t."

When we observe the initial sound of *pin* and the second sound of *spin* we think they are the same. Physically, in fact, they are quite different, as pointed out earlier. The *p* of *pin* is aspirated and is denoted phonetically as [pʰ]; the *p* of *spin* is not aspirated and is denoted as [p]. As a speaker of English you were probably unaware of the difference. If you were a speaker of Thai the [pʰ] of *pin* and the [p] of *spin* would sound as different as A from B.

If we say the word *pin* with the unaspirated "p" of *spin,* it would still be the word *pin* albeit with a "weird" pronunciation that might lead to confusion with the word *bin.* If we spit out the "p" in *spin* as an aspirated "p," it would still be the word *spin,* again with an unusual pronunciation.

Speakers of English all agree that there is one "p." The ordinary alphabet is accurate in this regard. Indeed, I'm sure I'm having trouble convincing some readers that physically, at least, there is more than one "p." Substituting the different "p's" doesn't produce a different word, it just alters the pronunciation. Psychologically, "p-ness" is preserved under aspiration. Substitute a "k" (change the place of articulation), however, and the word changes: *kin* is not *pin, skin* is not *spin.*

To account for these observations, linguists have posited the concept of the **phoneme.** Phonemes are the basic units in the sound system of a language. The phoneme is an abstraction required to study language systematically, but it also has psychological validity. This was observed nearly 60 years ago by the great American linguist Edward Sapir, who wrote:

> In the physical world the naive speaker and hearer actualize and are sensi-
> tive to sounds, but what they feel themselves to be pronouncing and hear-
> ing are "phonemes."

The feeling that speakers of English have that there is only one "p" in their language is a reflection of the fact that there is only one "p" **phoneme,** although English demonstrably has more than one kind of "p" sound.

Substituting one phoneme for another in a word changes that word, either into a different word, or a combination of sounds that isn't a word at all. For example, substituting /z/ for /s/ in *sink* gives a different word, *zinc.* The same substitution in *sock* would give the nonword *zock.* Substituting one pronunciation of a phoneme for a different pronunciation of the same phoneme produces the same word with an odd pronunciation.

Each abstract phoneme may be actualized—pronounced—in a variety of ways. The phoneme /p/—we denote phonemes between slashes, just as we write sounds between square brackets—has two actualizations, called **allophones.** The nonaspirated allophone exemplified in *spin* is denoted [p]. For the aspirated /p/ of *pin* we used a different symbol, because technically it is a different sound. We used [pʰ]. The phoneme /t/ has several allophones: [tʰ], [t], [D], [ʔ]. The schwa ə is an allophone of every English vowel phoneme, occurring in syllables lacking primary or secondary stress.

Phonemes are abstract. We think phonemes, but we speak allophones. Which allophone of a phoneme is pronounced in actual speaking depends on the **phonological rules** of the language. These rules are numerous and occasionally complex. They are not thoroughly understood by linguists, but some rules are well-known. We stated one informally in the previous paragraph when we described the schwa as an allophone of every vowel, occurring when the vowel is unstressed in the word.

The phonological rule that determines when the aspirated allophone of /p/ is realized is another example:

Pronounce /p/ as [p^h] when it occurs as the first consonant in a stressed syllable.

Thus the /p/ in *repel* is pronounced [p^h], because it is first in the stressed syllable *-pel.* The /p/ in *spin* is not aspirated, because it is not the first consonant. The /p/ in *apish* is not aspirated, because it is not in a stressed syllable (the stress is on the *a*). The same rule applies to the phonemes /t/, /k/, and /ch/, all of which have aspirated and unaspirated allophones.

The phonological rules are subconscious. We use them without thinking, as a manifestation of our linguistic knowledge. At our usual level of awareness, we don't notice allophonic variation, that the "p's" in *pin* and *spin* are different. We might notice, however, if someone deliberately switched the allophones and pronounced *spin* as [sp^{h}in] and *pin* as [pin]. We would also find it strange if a person or speech synthesizer used just one allophone of /p/ in both *pin* and *spin*. We might not know why, but we would certainly perceive that the speech had "an accent."

Additionally, awareness of phonemes allows us to ignore the infinite, minor variations in pronunciation, such as different degrees of aspiration, vowel length, or nasalization, that occur. That awareness also helps us to understand a person whose pronunciations are affected by illness and to recognize and understand the speech of many different persons (within limits—dialect variation often engenders allophone variation which may confuse us).

Every dialect of every language has its own inventory of phonemes along with the particular phonological rules that determine how a given phoneme is pronounced in a given context. Every language user in the world, whether speaking or listening, mentally links speech sounds to phonemes. Aggregates of phonemes form syllables and words, and their aggregates phrases and sentences.

Tables 1.11 and 1.12 present English consonant and vowel phonemes. They are similar to the tables of speech sounds given earlier—phonemes are still classified in the same way as the speech sounds that manifest them—without the various allophonic speech sounds such as voiceless aspirated stops, the glottal stop, the flap, and the schwa.

Table 1.11
English Consonant Phonemes

	Bilabial	Labio-Dental	Dental	Alveolar	Post-Alveolar	Palatal	Velar	Glottal
stop								
vcless	p			t			k	
voiced	b			d			g	
nasal stop	m			n			ng	
fricative								
vcless		f	th	s	sh			h
voiced		v	dh	z	zh			
approximant				l	r	y	w*	
affricate								
vcless						ch		
voiced						j		

*The *w*-phoneme in English is manifested with two approximate obstructions, one at the velum and one at the lips. It is customary to classify the phoneme as a velar approximant.

Table 1.12
English Vowel Phonemes

	Tongue Front	Tongue Central	Tongue Back
tongue high	ē (ai & oi end)		*ū* (ou end)
tongue high-mid	i		*u*
tongue mid	ā		*ō* (oi start)
tongue low-mid	e	ŭ	*ô*
tongue low	ă	a (ai & ou start)	o

Note: Diphthongs *ai* of *bite*, *ou* of *bout* and *oi* of *boil* are indicated by showing their tongue-start and tongue-end positions. *Boldfaced italicized vowels are rounded.*

In a sense, every phoneme has infinitely many allophones. That is because on the purely physical level, every pronunciation of /p/, say, is slightly different. Much of the variation is conditioned by the surrounding phonemes. The /p/ in *sample,* surrounded by /m/ and /l/ is slightly different than the /p/ in *open,* surrounded by two vowels. This phenomenon is known as **coarticulation.** Getting the coarticulation right, as native speakers do without thinking, is part of what makes speech sound natural. Not getting it right is part of what makes machine or synthetic speech sound unnatural (see a thorough discussion of this topic in Chapter 4). It's also part of what makes nonnative speakers speak with an "accent."

Some allophonic variation is random—we are not machines. (Likewise, machines aren't us, which is another reason why machine speech often sounds mechanical; ironically, it's too precise.) Some variation is socially conditioned. Some depends on our physical or emotional state. Some depends on whether a man, woman, or child is speaking. Some depends on whether the speaker is a native speaker of English. Some depends on the region of the country where the person grew up. Some is systematic, such as the rule governing the occurrence of [p] and [p^h]. Without the phoneme, language would be impossible. We would never know when a difference was significant, as that between [p] and [f], or insignificant, as that between [p] and [p^h]. Phonemes, together with the phonological rules, are how we deal with the infinite variation in the pronunciation of speech sounds.

Despite the impossibility of a "phonetic" alphabet that would reflect every minute difference among speech sounds, it is possible to compromise and design a phonetic alphabet where the level of detail has significance. The IPA is an outstanding example. Its design adheres to the following principle: If in some language a difference in some feature of pronunciation can be used to make phonemes—and hence, words—different, then include phonetic symbols that reflect such a feature. If a phonetic difference is never used to distinguish phonemes, do not include phonetic symbols for that difference.

For example, you can utter /p/ with your teeth clenched. Denote this sound as p^c. It will sound noticeably different than any allophone of /p/ uttered without teeth clenching. Should we include the symbol p^c in a phonetic alphabet, and should we consider [p^c] an allophone of /p/ in English? The answer is "no" because no language of the world uses teeth clenching as a feature for distinguishing phonemes.

In Thai, on the other hand, aspiration is used to make different phonemes, and, hence, different words. The word pronounced [pa] means "forest." If the "p" is aspirated to give [p^{h}a], you get a different word that means "to split." Thus the IPA has different phonetic alphabet symbols for distinguishing aspirated versus unaspirated consonants.

When we analyze English we distinguish [p] and [pʰ] as allophones of /p/, but we ignore the hypothetical allophone [pᶜ]. When we analyze Thai we find there are two phonemes, /p/ and /pʰ/, with allophones [p] and [pʰ], respectively (among others).

A truly phonetic alphabet of English, even at the IPA level of detail, would be unwieldy. All English vowels have nasal allophones, for example. The /i/ in *pin* is phonetically [pʰi˜n] where the superscript tilde following the *i* means nasalized (i.e., air is allowed to flow out both the nose and mouth). Nasalization in English, like aspiration, is not phonemic. We don't make different words by nasalizing vowels. In French, however, nasalization is phonemic. The word [bo] *beau* and the word [bo˜] *bon* mean different things. So according to the principle, an IPA analysis of English must reflect nasalization. That doubles the number of vowel symbols in one stroke. Each vowel has a nasal allophone that occurs whenever it shares a syllable with a nasal consonant. Otherwise, the non-nasal (oral) allophone is found.

There are many similar cases that, if accounted for, would lead us to an English phonetic alphabet of well over one hundred different symbols.

Such an alphabet would not be practical for speakers of English. It would contain more detail than was needed to pronounce words. I don't have to spell *pit* [pʰit] to get you to pronounce the aspiration. The phonological rules of English literally force you to pronounce it that way. On the other hand, if I am designing a speech synthesizer for English, or a segment-based speech recognizer for English, I had better know about allophones or my synthesizer will sound unnatural and my recognizer won't work.

1.5 Articulatory Processes

Speak the speech, I pray you . . . trippingly on the tongue.

William Shakespeare, *Hamlet*

I was in my office the other day at about 12:45, too busy for lunch at noon. A colleague whom I sometimes go to lunch with poked his head in and said "[jēcheʔ]." Say what? If I tried to spell this obscure remark it would come out "geechet" though that wouldn't reflect the glottal stop, which has to be spelled with a "t." Was I baffled by these two syllables? The answer is not at all, and neither would you have been in the same situation. My answer was, "No, how 'bout McDonalds?"

If my friend had articulated carefully it would have come out as [didyuetyet]. Had he been slightly more informal than that, but still more careful than what he actually said, it might have been [dijəēchet] or even [jəēchet].

What we are seeing are the articulatory effects of the running together of syllables and words. Most speech is run together. We are able to understand it, even transcribe it fully spelled out, because of our knowledge of the language, which, together with context and past experience, help us to figure out what people are saying when they do not speak "trippingly on the tongue." Given the time of day and an utterance by my occasional lunch companion, I knew instantly that he was asking, "Did you eat yet?"

Lyricists never tire of playing with articulatory phenomena. A song of the 1940s entitled *Mairzy Doats* ("Mares eat oats") and the Gerschwins' (George and Ira) famous *S'Wonderful* are but two examples.

The degradation of speech at syllable boundaries is for the most part systematic, or rule-governed, just as it is for most aspects of speech such as allophone distribution. If the degradation were random, communication would break down. The slurred speech of an intoxicated person is more difficult to understand because some of the articulation is random under the neurological numbing of alcohol or drugs. Similarly, neurologically disturbed speech, termed **dysarthria,** which may result from disease or injury, sounds garbled because the deviations generally don't follow the rules with which listeners are familiar.

The degree of degradation of "normal" speech has to do with the informality of the speech and the rate at which it's spoken. There is some individual variation, to be sure. We all know people who "talk fast," or conversely, "talk slow" and enunciate. Nevertheless, most articulatory effects are rule-governed and predictable. Let's take a brief look at a couple of well understood rules.

Starting with "did you eat yet," there is a rule that says when a [d] and a [y] come together, the result is a [j]. Linguists call this process **palatalization.** The alveolar [d] anticipates the palatal articulation of the [y]. The tongue never completes its trip to the alveolar ridge. It gets only as far as the palatal area and out comes the palatal affricate [j]. This is an example of a more general process called **anticipatory coarticulation.** The identical phenomenon collapses the [t] and the [y] into the voiceless affricate [ch]. From palatalization alone, "did you eat yet" becomes [dijuechet].

We already know that every vowel phoneme has the allophone [ə] when it is unstressed. This rule reduces the ū to a schwa giving [dijəēchet].

Language has many rules of **elision.** The word comes from the verb *elide* which is linguistic jargon for "to omit." English has a rule in which the first syllable of a word or phrase may be elided, provided it does not have primary stress. It's where the nickname "Lizbeth" for "Elizabeth" comes from, and why we say *'bout* for *about.* Also, in rapid speech, the schwa may be elided. Now we've got [jēchet]. The glottal stop is an allophone of /t/ when it occurs at the end of a syllable, especially in rapid, informal speech, giving us, finally, [jēche?].

To a computer attempting to understand rapidly spoken speech, such articulatory effects are extremely difficult. Without heavy linguistic and contextual knowledge, it is nearly impossible to reconstruct [didyuetyet] from [jēche?]. This is the great challenge to computer recognition of continuously spoken speech, which we will return to in Chapter 3.

People have their problems with this phenomenon, too—including myself: For years I took the phrase "for all intents and purposes" as "for all intensive purposes." Don't ask me why. I suppose I was hearing [intensnpurpəsəz] and confused the second [n] with [v] in attempting to make sense out of the idiom. Everyone (including computers) does such things, and these "errors" are a common source of embarrassment and humor. When I am invited to give a talk on speech recognition, I call my talk "How To Wreck a Nice Beach." If the word gets around a few environmentalists show up looking for a fight. When they discover my joke, that I really mean "how to recognize speech," they usually leave.

References

[1] Pinker, S., *The Language Instinct: How the Mind Creates Language*, New York: W. Morrow and Company, 1993.

[2] Ladefoged, P., *A Course in Phonetics, 3rd Edition*, Fort Worth, TX: Harcourt Brace College Publishers, 1993.

[3] International Phonetic Association, *The Principles of the International Phonetic Association*, London: Department of Phonetics and Linguistics, University College London, 1949 (latest revision 1989).

[4] Pullum, G. K., and W. A. Ladusaw, *Phonetic Symbol Guide*, Chicago: The University of Chicago Press, 1986.

[5] Fromkin, V., and R. Rodman, *An Introduction to Language, 6th Edition*, Fort Worth, TX: Harcourt Brace College Publishers, 1998.

[6] Borden, G. J., K.S. Harris, and L.J. Raphael, *Speech Science Primer: Physiology, Acoustics, and Perception of Speech, 3rd Edition*, Baltimore, MD: Williams & Wilkins, 1994.

[7] Lass, N. J., (ed.) *Principles of Experimental Phonetics*, St. Louis, MO.: Mosby, 1996.

2

Representing Speech in the Computer

The notion of a *representation* of a speech signal is central to almost every area of speech communication research.

Ronald W. Schafer and Lawrence R. Rabiner, *Digital Representations of Speech Signals*

2.1 Introduction

When we hear speech we perceive subconsciously many of the frequency components of the speech signal. This is demonstrated by the fact that we hear differences among vowel and consonant sounds. We distinguish between [a] and [ē] without conscious knowledge of formant frequencies, just as we distinguish the colors red and blue without conscious knowledge of their wavelengths.

The human auditory system evolved specific mechanisms for processing speech. The ear is especially sensitive to the range of frequencies that compose speech; it responds to them by varying the neural signals sent to the brain. It's a bit like a mechanical coin sorter at the bank. When you empty your piggy bank into a coin sorter, it separates the pennies, nickels, dimes, and quarters into individual slots in a tray. You can compare the relative amounts of the various coins quite a bit more easily then. The auditory system sorts the different frequencies that constitute sound into their relative amplitudes for the brain to process.

Computers process speech in a similar way. The computer's "ear" consists of a microphone and an **analog-to-digital (A-to-D) converter.** The microphone converts the vibrations of sound to an electrical signal. The A-to-D converter reduces the continuously varying voltages of the electrical signal to a sequence of digits. Computer programs decompose the digitized signal into its constituent frequencies, which may be processed further to determine, for example, which vowel was spoken based on the location of formants.

At each stage of computer processing, inaccuracies occur and accumulate. In this chapter, we will examine the processing methods and the deviations from the actual signal that are introduced.

2.2 Microphones

Inside every microphone is a diaphragm, capable of vibrating in concert with any sound whose frequencies are within its range of operation. These oscillations are converted into electrical signals in a variety of ways depending on the type of microphone. In a carbon microphone, often found in telephones, the level of resistance in an electrical circuit is controlled by the oscillations so that a variation in electrical output replicates the original sound. In a condenser microphone a variation in capacitance produces the same effect. Vibration induced variations of electromagnetic fields and shapes of piezoelectric crystals are also used to control the transducing of sound to an electrical signal.

Microphones are designed for various patterns of reception and various placements in the environment. **Omni-directional microphones** collect sounds from all directions. **Uni- and bi-directional microphones** have maximum sensitivity to sound coming from one or two directions. Microphones may be handheld, the favorite of rock singers; attached to the lapel, the favorite of talk show guests; head-worn, the favorite of telephone operators; hung from tall ceilings, the favorite of concert pianists; or stuck in the ear, nobody's favorite.

Noise-canceling microphones are important when noise is not well tolerated, as in computer speech recognition. A typical noise-canceling microphone is actually two microphones, one directed at the speaker and the other in the opposite direction. Ambient noise enters both microphones at about equal levels of amplitude, but the amplitude level of speech is much higher in the speaker directed microphone. Signals common to both microphones are subtracted out, leaving mostly speech signal, which is then amplified and transmitted. This is illustrated in Figure 2.1.

Microphones nowadays may be wireless. Their electrical output is transmitted as an electromagnetic wave to a receiver. Wireless mikes are becoming increasingly popular as their fidelity improves with technological advances.

Generally, neither microphones nor ears capture all of the information in a signal when transducing its mechanical vibrations into an electrical signal. The human ear, of course, is insensitive to sounds outside certain frequency and amplitude ranges. Microphones, depending on their design, may also not fully

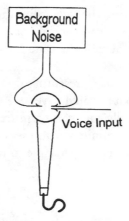

Figure 2.1　Noise-canceling microphone. The speech signal enters the microphone primarily through the opening nearest to the source. Background noise enters both openings, allowing the circuitry in the microphone to subtract it out from the combined signal and noise input.

capture the frequency and amplitude fluctuations of a signal with a full degree of fidelity. Thus, from the start, speech processing, whether by computer or human, suffers a certain amount of information loss.

2.3 Sampling

The electrical output of a microphone is an analog signal. It varies smoothly with time. To every point in time (the *x*-axis) it has an amplitude (the *y*-axis) that corresponds to its voltage. Figure 2.2 is an illustration.

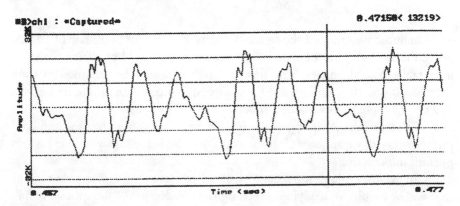

Figure 2.2　An analog signal.

In Figure 2.2, the vertical line identifies the point time = 0.47150, amplitude = 13,219 (shown in the upper right hand corner). Every time value has a corresponding amplitude value—there are no gaps. That is what is meant by **analog**.

Consider the problem of representing this function without recourse to the figure, the way a computer would have to. All that can be done is to represent *some* of the points, ideally enough to capture the vital characteristics of the signal. An analog signal has infinitely many x-values to which y-values must be assigned, so that it is impossible in principle to represent it value-by-value in a computer.

2.3.1 Sampling Rate

Since analog signals have infinitely many points, the best a computer can do is keep track of finitely many of them and hope they are a representative sample. Most common is to sample the signal at equal time intervals, noting at each time the corresponding amplitude. Physically, the sampling is achieved by an A-to-D converter.

The **sampling rate** is the number of points looked at each second. A sampling rate of 1,000 would record values 1,000 times each second, or once a millisecond (ms). The **sampling period** is the amount of time that elapses between each sample. With a sampling rate of 1,000, the sampling period would be 1/1,000 of a second, or 1 millisecond. With a sampling rate of 10,000 the sampling period would be 0.1 of a ms. The sampling period is the inverse of the sampling rate and vice versa. In mathematical terms, if *SP* is the sampling period and *SR* is the sampling rate, then

$$SP = \frac{1}{SR} \qquad \text{and} \qquad SR = \frac{1}{SP} \qquad (2.1)$$

Suppose we sampled the signal in Figure 2.2 at a sampling rate of 10,000. Since the total amount of time is 0.02 seconds, we would represent the signal in a table of 200 entries. The first few points would be recorded at times 0.45700, 0.45710, 0.45720, . . . The sampling period is the difference between any two of these points, namely 0.0001 seconds, or 0.1 ms.

Choosing the sampling rate is important. If it's too high, time and computer storage space are wasted. If the sampling rate is too low, you may lose important information.

To examine the ramifications of sampling, let's look at a simple example that will be helpful in studying speech signals (see Figure 2.3). Suppose we have a sine wave—an analog signal identical to the trigonometric sine function.

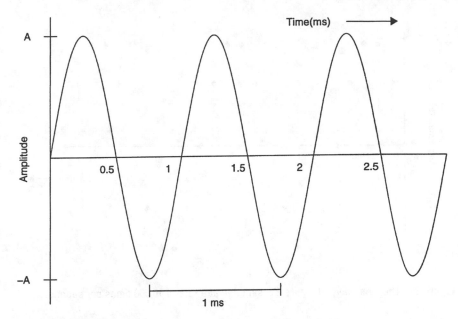

Figure 2.3 Sine wave with amplitude *A*, frequency 1,000 Hz, period 1 ms.

A sine wave is a **periodic function,** one that repeats over time. The time between repetitions is called the **period.** The **frequency** is the number of repetitions per second, measured in hertz. The *y*-value for any given *x*-value is the **amplitude** of the sine wave at that point in time. In Figure 2.3, the sine wave has a period of 1 ms and a frequency of 1,000 Hz. The maximum amplitude reached, usually called **the amplitude of the sine wave,** or simply "the amplitude," is the *y*-value *A* in the illustration. This sine wave represents a pure tone corresponding to a musical note roughly two octaves above the middle C of a piano keyboard.

The frequency and the period are inverses of each other. In mathematical terms, if *P* is the period and *F* is the frequency, then

$$P = \frac{1}{F} \quad \text{and} \quad F = \frac{1}{P} \tag{2.2}$$

If we sampled this sine wave at a sampling rate of 10,000 times per second, starting at the beginning of a period, we would have 10 points for each period. After carrying out this A-to-D conversion, the smooth curve of Figure 2.3 would look like the illustration in Figure 2.4. This is actually what the computer would "see."

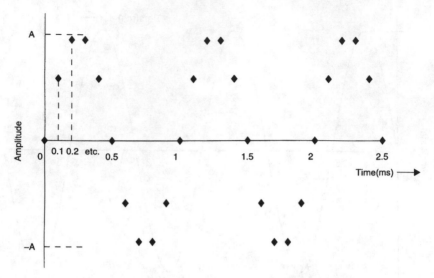

Figure 2.4 The sine wave of Figure 2.3, sampled at a rate of 10,000 times per second.

Is the sampling rate sufficient? To answer this question we must pose a different question: Using a reverse of the A-to-D process, called **Digital-to-Analog (D-to-A)** conversion, can we reconstruct the original analog signal from the abbreviated digital representation? If we pretend we don't know that the original signal is a sine wave, the answer is "no." We don't know what to do between points. We can interpolate, most simply with straight lines, and while some of the character of the sine wave is present, some precision has been lost, as can be seen by comparing Figure 2.5 with Figure 2.4.

If we knew in advance that the sampled curve was a sine wave of amplitude *A,* we could interpolate with 100% accuracy. Ten samples per period is more than enough to reconstruct a sine wave. How few samples can we get away with and still recover the original wave without error? The answer is surprisingly small:

> Fact 1: Two or more samples per period are sufficient to reconstruct a sine wave of known amplitude.

As an illustration, suppose we have sampled a sine wave 2,000 times per second, so we have samples at 0.5 ms, 1.0 ms, 1.5, ms, etc. To keep it simple, we may assume, without loss of generality that the samples are taken at the times where the sine wave is zero (called the **zero-crossings**). Figure 2.6 represents a few sampled points.

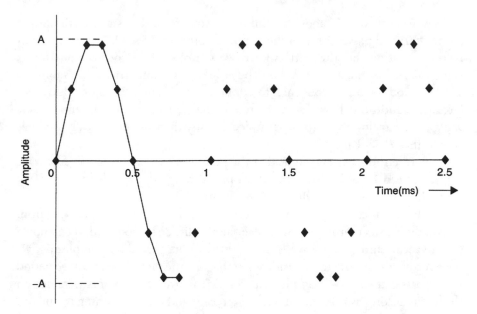

Figure 2.5 The sampled sine wave of Figure 2.4 partially reconstructed without prior knowledge that it is a sine wave.

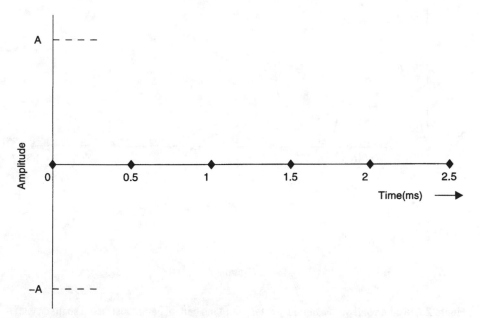

Figure 2.6 A sine wave sampled twice per ms (sampling rate = 2,000 samples per second), with samples lying on the *x*-axis.

What can we say about an unknown sine wave that starts at the origin (time = 0) and passes through the sampled points at 0.5, 1.0, 1.5, etc., subject to the condition that there are at least two samples per period? For one thing, the period of such a sine wave must be at least 1 ms long. If it were shorter, a single period would be too small to contain two or more samples, which violates the condition. If we let P_x represent the period of some unknown sine wave we are trying to fit to the sampled points, then mathematically we have shown that $P_x \geq 1$ ms.

Furthermore, we can argue that the period of such an unknown sine wave cannot exceed 1 ms. If it did, it couldn't pass through the sampled point at 0.5; it would "overshoot" it, as illustrated in Figure 2.7.

Mathematically, $P_x \leq 1$ ms. Together with the inequality $P_x \geq 1$ ms from above, this shows that $P_x = 1$ ms. Since the amplitude is presumed to be known (in advance), and a sine wave is determined by its period and amplitude, we have reconstructed the unique sine wave that fits the samples and whose period encompasses at least two samples. (It wouldn't matter if the points were not on the x-axis as long as there are at least two per period—the reasoning is similar.)

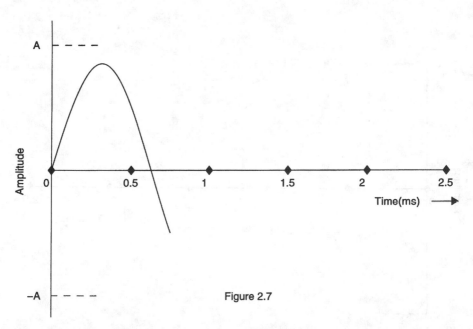

Figure 2.7

Figure 2.7 A sine wave with a period greater than 1 ms cannot be reconstructed so that it will fit the samples taken from a sine wave whose period is exactly 1 ms. It misses the point (0.5, 0).

Without the condition that there are at least two samples per period, or, equivalently in our illustration, that the reconstituted sine wave has a frequency less than or equal to 1,000 Hz, the uniqueness of the reconstruction is lost. The sample can be fit by infinitely many sine waves, each of whose frequencies is a multiple of the frequency of the sine wave of lowest frequency that fits the sampled points. This is illustrated in Figure 2.8.

This is an unsatisfactory situation. The reconstruction must be unique. Therefore, in reconstructing continuous sine waves from sampled sine waves, we reconstruct the one with the largest possible period, or smallest possible frequency, that fits the sampled points.

This brings us to the subject of inadequate sampling rates, or **undersampling.** Suppose we sample a 1,000 Hz sine wave at a rate of 1,000 samples per second, or only one sample per 1-ms period. When we reconstruct the sine wave of largest period that fits the samples, we get a sine wave whose period is two milliseconds, and whose frequency is 500 Hz, half that of the original sine wave. This is illustrated in Figure 2.9.

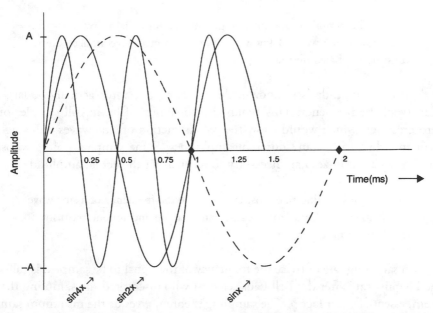

Figure 2.8 Sine waves of one-half and one-quarter of the period of the sine wave with the lowest frequency or largest period (dashed curve) that fits the sampled points.

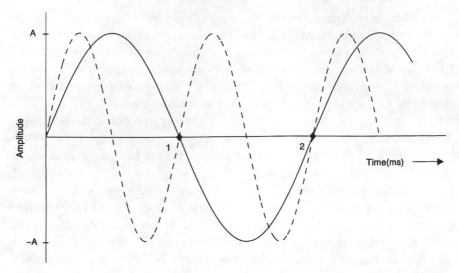

Figure 2.9 Fitting the sine wave of largest period to points sampled from a 1,000-Hz sine wave at a sampling rate of 1,000. Its frequency is 500 Hz or a period of 2 ms. The originally sampled sine wave is the dashed curve.

Based on fact 1 and the discussion of under-sampling, we arrive at fact 2.

Fact 2: The sampling rate must be greater than or equal to twice the frequency of a sine wave of known amplitude to guarantee an accurate reconstruction based on the samples.

If the amplitude is unknown, the sampling rate must actually be larger than twice the frequency. This avoids the possibility of having all samples on zero crossings, which would allow the reconstruction of sine waves of a single frequency, but with many different amplitudes. Some sampled points whose amplitudes are non-zero are necessary to reconstruct the actual amplitude.

Fact 3: The sampling rate must exceed twice the frequency of a sine wave of unknown amplitude to guarantee an accurate and unique reconstruction based on the samples.

A sampling rate of twice the frequency of the signal to be sampled is called the **Nyquist rate** after the Bell Labs scientist who discovered **the sampling theorem,** essentially our fact 3. The sampling theorem gives us the minimum sampling rate for the accurate reconstruction of a sine wave. Previous discussion explains why under-sampling leads to the reconstruction of a sine wave of lower frequency than the one that was originally sampled. The circumstance that results from under-sampling is called **aliasing.**

> Fact 4: Aliasing occurs whenever the sampling rate is less than the Nyquist rate and results in higher frequencies being falsely reconstructed as lower frequencies.

An illustration of the effect of aliasing in music would be as follows. Suppose you are sampling a musical note whose pitch (frequency) is 440 Hz. (That is the note "A" to which orchestras tune their instruments.) If you sampled at a rate less than 880 times per second, when you reconstructed the note based on the samples you would actually hear a note of pitch lower than that of the original "A."

Aliasing occurs in functions other than signal sampling. A movie camera is a sampling device. It samples its subject as individual frames, taking enough frames per second to ensure that the eye and brain will reconstruct a smooth scene. If you have a videocassette player with a "freeze" and a "step" function, you can watch your favorite movie frame by frame. It is surprising how disjointed adjoining frames are.

When the camera is sampling a periodic phenomenon whose frequency exceeds half the sampling or frame rate of the camera, you get aliasing, or a false impression of the scene. This is why spinning airplane propellers appear incongruous in the movies. They either seem to be motionless, or moving backward, or moving forward at an unrealistic rate.

Let's make up some numbers to see how this might work. Suppose the propeller is rotating clockwise at a frequency of 21 revolutions per second and our camera snaps 24 frames per second, well below twice the frequency. Suppose at frame one the propeller is in the position shown in Figure 2.10.

The next frame is 1/24th of a second later. The propeller moves 21/24 of a full circle, and appears in frame two as shown in Figure 2.11.

Figure 2.10 First movie frame of a rotating propeller.

Figure 2.11 Second movie frame of a rotating propeller.

In frames three, four, and five, the propeller appears as shown in Figure 2.12, advancing each frame by 21/24 of a full circle from its previous position.

What does the eye see when the frames are played back? It sees a propeller that appears to be rotating backwards at the lazy rate of three times per second. No wonder James Bond finds it so easy to outrun movie airplanes!

2.3.2 Quantization

Sampling at too low a rate results in errors in the form of aliasing when the signal is reconstructed. Aliasing, however, is not the only source of potential error

Figure 2.12 Third, fourth, and fifth movie frames of a rotating propeller.

in reconstructing signals based on sampling. There is also the problem of **precision.** The largest and most powerful digital computer in the universe shares a limitation with the abacus, the electric adding machine, the handheld calculator, your fingers and toes: none can represent most numbers precisely.

Divide one by three on your calculator. You'll most likely see 0.3333...3, the number of threes depending on the calculator. This is not really 1/3; it's an approximation. No matter how big a calculator you buy, no matter how many threes you string out, it will still be an approximation.

A computer has to represent numbers using only bits (i.e., zeros and ones). Because of this, most numbers are approximated in a computer. By using copious quantities of bits these approximations can be made better and better. That is, the error, the difference between the actual number (e.g., 1/3), and the approximation (0.3333. . .), can be made as small as desired, but it never goes away entirely.

Quantization is the process of representing a range of numbers with a fixed number of digits, rounding to the nearest representable value where necessary. In computers, it is the process of representing a range of numbers with a fixed number of bits (recall that *bit, from binary digit,* is one of the two digits, 0 or 1). The device that performs quantization is called a **quantizer.** The number of bits used by the quantizer is termed the **resolution** of the quantizer. A quantizer of *B* bits, called a B-bit quantizer, is capable of representing 2^B distinct values, called **quantization levels.**[1]

To be more concrete, suppose we sample a sine wave at 10,000 samples per second. Then, for each second of signal duration we need to store 10,000 amplitude values. Most of these values cannot be represented precisely in a computer regardless of how many bits are dedicated to the task. Worse, though, is that the number of bits that can be used to represent each value is necessarily limited.

If we use a 10-bit quantizer—giving us about three decimal digits of precision—we would require 100,000 bits or 12,500 bytes (eight bits to the byte) per second of signal. In practice, we may want to store lengthy signals. One hour's worth is 45 megabytes of storage, a sizable bite out of any hard drive. A 1.44-megabyte diskette can barely store a minute of speech represented this way. Transmission of speech in this form puts a large demand on communication lines. Thus there are good reasons to limit the size of the quantizer. In practice, quantizers rarely exceed 16 bits of resolution.

---

1. It is not uncommon in the literature for the term resolution to be applied to the quantization levels. Thus, what I and other authors call B-bits of resolution may be called a resolution of 2^B. The confusion and oft-occurring ambiguity is due to the obvious equivalence between the B-bits used for representation and the 2^B levels of representation that result.

The ramifications of using a 10-bit quantizer are actually worse than you might think at first blush. Ten bits can only distinguish among 1,024 (2^{10}) values or quantization levels. These values may be the integers 0–1,023, or fractions between zero and one (1/1,024, 1/1,023, . . . 1/1), or whatever 1,024 values we wish. However, the 1,024 is an absolute limit.

Suppose the sine wave has an amplitude of 511. Its range is between a minimum of −511 and a maximum of 511. The amplitude values may be any numbers in that interval. With only ten bits per sample, we have to choose 1,024 representative values. Suppose we choose those values to be −512, −511, . . . 0, 1, . . . 511. Then the **quantizer step size**—the difference between quantization levels—is 1.0. How do we represent the amplitude 361.332? We must choose the nearest integral value, namely 361, introducing a **quantization error** of 0.332 or 0.09%. Most sampled values fall between quantization levels, resulting in an error that varies between 0 and 0.5, and averages 0.25.

When the signal is reconstructed, the accumulated quantization error will result in **quantization noise**. (**Noise** in this context is defined as unwanted deviations from the original signal.) By using more bits per sample, quantization error is reduced, but at the cost of increased storage requirements.

Quantization noise may be measured in decibels as a **signal-to-noise ratio** (SNR). In such a case the SNR is a measure of the relative, cumulative error in a reconstructed signal. The smaller the error, the larger the SNR, and the more faithful the reconstruction. It follows that the SNR can be increased by increasing the size of the quantizer. This fact may be quantified by a formula for the SNR that applies when the quantization levels are chosen so that the maximum level approximates the maximum signal amplitude, and the minimum level approximates the minimum amplitude. In other words, the amplitude range and quantization level range nearly match. (See [1], pp. 184–185, for a complete derivation of (2.3)). For a B-bit quantizer,

$$\text{SNR (in dB)} = 6B - 7.2 \qquad (2.3)$$

It follows from (2.3) that adding one bit to the quantizer improves the SNR by 6 dB. We observed in Chapter 1 that the relative intensity of a signal in dB is equal to $20 \log (P/P_R)$, where P_R is a reference level sound pressure and P is the sound pressure of the signal itself. For measuring SNR in decibels, we take the reference level to be the sound pressure level of the noise. Thus for a signal S, the SNR is $20 \log (P_S/P_N)$, where P_S and P_N are the sound pressure levels of the signal S and the noise N.

By adding one bit to the quantizer we double its resolution, so the average error is halved, which has the effect of halving the relative intensity of the noise P_N. The 6-dB improvement from (2.3) can be thought of as equivalent to this halving of the relative intensity of the noise P_N. This has the effect doubling the ratio of P_S / P_N, which, as you may recall from Chapter 1, does indeed correspond to a 6-dB increase.

The sampling rate and the quantizer size are the **parameters of sampling,** the two factors that ultimately determine how faithfully the original signal can be recovered after the A-to-D and D-to-A processes have occurred. The two parameters are independent of each other insofar as the fidelity of the digitized signal is concerned. An increase in the sampling rate will not reduce the overall quantization error, and aliasing will occur if the sampling rate is too low regardless of the size of the quantizer.

2.3.2.1 Nonlinear Quantizers

We have been dealing with pure sine waves whose amplitude remains constant throughout, and whose sampled values are equally dispersed along the y-axis. A linear quantizer, one whose values are equally spaced, is suitable for those cases. The range of quantization levels is chosen to match the amplitude range of the sine wave as closely as possible to make the most efficient use of the levels available. This has the effect of minimizing the quantization error for that particular quantizer. However, what if the amplitude is not constant?

A common situation is where relatively few values have very high amplitudes compared with the majority of points that have relatively low amplitudes. An example is the damped sine wave of Figure 2.13.

Most of the values of this sine wave are between 1 and 3, though we need to represent a few values as large as 7. Suppose we have a 4-bit quantizer, capable of representing 16 values. Linear quantizer levels might reasonably be:

0, 0.5, 1, 1.5, 2, 2.5, 3, 3.5, 4, 4.5, 5, 5.5, 6, 6.5, 7, 7.5

In particular, there are five values covering the range from 1 to 3, where most of the signal lies, and nine values covering the range from 3.5 to 7.5. Referring to Figure 2.13, the quantizer noise from each of the few sampled values in the range 3 to 7 (where t is less than one) will be the same as the quantizer noise from each of the many sampled values closer to zero that occur when t is greater than one. The error of samples varies from 0 to 0.25 with an average of 0.125.

A nonlinear quantifier would concentrate the greater part of its quantization levels in the lower range, achieving better resolution there at the expense of poorer resolution for the high values. However, since the high values are relatively few, the overall quantization error is reduced.

A common nonlinear quantizer is a **logarithmic quantizer.** Let's design one for Figure 2.13. We first replace each sampled value of the sine wave with its base 2 logarithm. (For example, if the value was 2, it would be replaced by

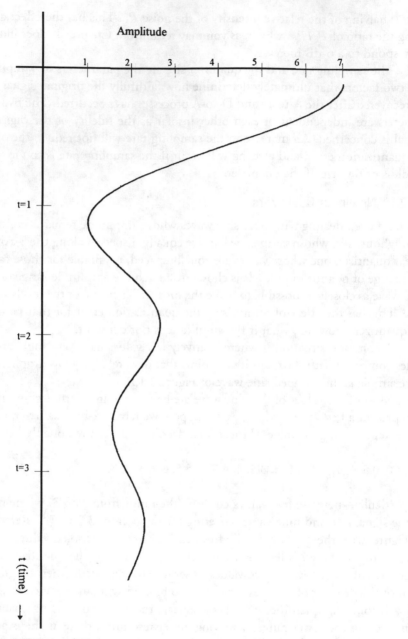

Figure 2.13 Damped sine wave.

1; if it was 4, it would be replaced by 2; and in general if it was x it would be replaced by $\log_2(x)$. The amplitude range of the log of the sine wave now 0–2.8 (i.e., $\log_2(1)$–$\log_2(7)$), as illustrated in Figure 2.14.

We now apply a linear quantizer to the transformed signal, letting the 16 values of the quantizer levels be distributed between 0 and 3:

0, 0.2, 0.4, 0.6, 0.8, 1, 1.2, 1.4, 1.6, 1.8, 2, 2.2, 2.4, 2.6, 2.8, 3.0

This is a linear quantizer on a logarithmic function. We now take the exponent of each of the above values (i.e., the anti-log) to produce the logarithmic quantizer on the original sine wave, as illustrated in the first two entries of the fourth column of Table 2.1, which summarizes all of this mathematical activity.

Now there are nine values covering the range between 1 and 3 and six values covering the range between 3.5 and 7. The average quantizer error in the

Table 2.1
Table for a Logarithmic Quantizer

Quantization Level Number	Linear Quantization Levels of Original, Damped Sine Function	Linear Quantizer Levels on Log of Original, Damped Sine Function	Logarithmic Quantizer Levels on Original, Damped Sine Function
1	0	0	1 (2^0)
2	0.5	0.2	1.148698 ($2^{0.2}$)
3	1.0	0.4	1.319508
4	1.5	0.6	1.515717
5	2.0	0.8	1.741101
6	2.5	1	2
7	3.0	1.2	2.297397
8	3.5	1.4	2.639016
9	4.0	1.6	3.031433
10	4.5	1.8	3.482202
11	5.0	2	4
12	5.5	2.2	4.594793
13	6.0	2.4	5.278032
14	6.5	2.6	6.062866
15	7.0	2.8	6.964405
16	7.5	3	8

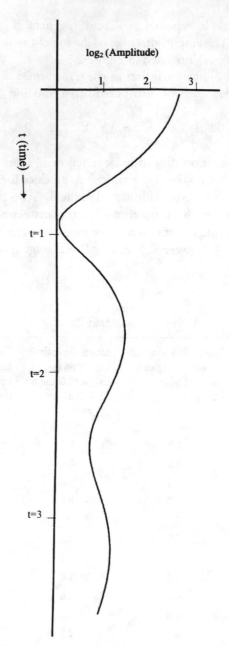

Figure 2.14   Log of damped sine wave of Figure 2.13.

1–3 range is approximately 0.07, and in the 3–7 range approximately 0.19, but since many more values are in the lower range, the overall quantizer error introduced by the logarithmic quantization is less than the quantizer error introduced by a straight linear quantization.

Linear and logarithmic quantizers are the most common because they are easy to implement and generally effective. **Adaptive quantizers** change their range, and hence their resolution, dynamically, adjusting to the signal as its range changes. In the damped sine wave, an adaptive quantizer might detect that in the time interval 0–1, the range of the signal is considerably greater than in the time interval 1–2 and readjust its levels accordingly for the interval 2–3.

The physical processes of sampling and quantization are done by the electronic components that I have referred to as A-to-D and D-to-A converters. Often, these two devices are combined in a single integrated circuit. They are found in all digital telephones and are included in sound cards for PCs, as components in work stations, and in other environments where sound is digitally processed.

2.4 Speech Digitization

Here is what we've learned so far: To represent an analog or continuous signal as a stream of numbers we must sample it at discrete times, and quantize the value we get to the nearest quantization level. How good a representation we get depends on the **sampling rate**, which must be high enough to avoid aliasing, and the **resolution of the quantizer**—the number of bits used to represent each value.

Sine waves have been used to illustrate these concepts in part because they are familiar to most readers. However, there is another reason. All periodic functions, no matter how complex, can be represented as sums and differences of sinusoidal (sine and cosine) waves of varying periods, amplitudes and phases.[2] These are called Fourier series after the French mathematician Jean B. J. Fourier (1768–1830) who first used them.

Here are several of the mathematical expressions pertinent to Fourier analysis, presented for the benefit of the mathematically curious. Books at all levels have been written on Fourier analysis and its applications. The interested reader may consult any number of them, for example, [1].

---

2. A cosine wave is identical to a sine wave displaced 1/4 of a period to the left so at the *y*-axis (zero time), it obtains its greatest positive value. Thus, everything we've said about sine waves applies equally to cosine waves.

A periodic signal $F(t)$, with period T, has the characteristic that $F(t) = F(t + kT)$ for values $k = 0, 1, 2, \ldots$ In words, the values of the function F repeat every T milliseconds (or whatever the unit of time happens to be). Fourier analysis tells us that such a signal may be approximated as a sum of sine and cosine functions, namely

$$F(t) \approx a_0 + a_1\cos\left(\frac{2\pi t}{T}\right) + b_1\sin\left(\frac{2\pi t}{T}\right) + a_2\cos\left(\frac{4\pi t}{T}\right) +$$

$$b_2\sin\left(\frac{4\pi t}{T}\right) + a_3\cos\left(\frac{6\pi t}{T}\right) + b_3\sin\left(\frac{6\pi t}{T}\right) + \ldots$$

The approximation error can be made as small as desired by including more and more terms. A summation sign on the index n can be used to express the formula concisely

$$F(t) = a_0 + \sum_{n=1}^{\infty}\left(a_n\cos\left(\frac{2\pi nt}{T}\right) + b_n\sin\left(\frac{2\pi nt}{T}\right)\right) \qquad (2.4)$$

For every value of t that is a multiple of T, the cosines and sines "start over," capturing the periodic nature of F. Successive values of n produce cosines and sines of increasing frequency, with each increase a multiple of the fundamental frequency $1/T$. These are the **harmonics** of the periodic function. This means of representing a periodic function is often called **harmonic analysis**. Figure 2.15 is a simple illustration.

The a-coefficients and b-coefficients are the amplitudes of the cosine and sine components. They are important because they describe how much each harmonic contributes to the overall signal. Their computation takes us momentarily into the realm of integral calculus:

$$a_0 = \frac{1}{T}\int_0^T F(t)\,dt$$

$$a_n = \frac{2}{T}\int_0^T F(t)\cos\left(\frac{2\pi nt}{T}\right)dt \qquad n \geq 1 \qquad (2.5)$$

$$b_n = \frac{2}{T}\int_0^T F(t)\sin\left(\frac{2\pi nt}{T}\right)dt \qquad n \geq 1$$

These integrals may be computed providing the values of $F(t)$ are known to the computer in the interval of 0 to T. Sines and cosines may be expressed in com-

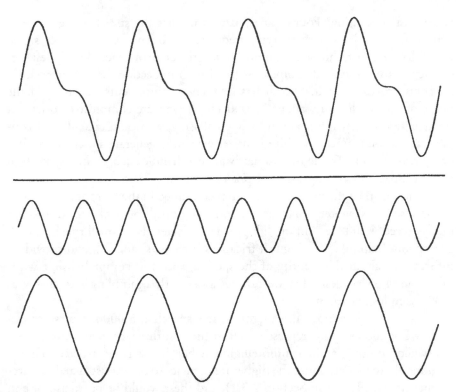

Figure 2.15 The complex periodic wave above the line may be decomposed into the sum of the two sinusoidals below the line.

plex exponential form, and therefore so can equation (2.4). This is given as (2.6). (j is the imaginary unit; see [2] for a derivation.)

$$F(t) = a_0 + \sum_{n=1}^{\infty} c_n e^{\frac{2\pi jnt}{T}} \quad \text{where } c_n = a_n - jb_n \qquad (2.6)$$

Many books on signal processing use this form. Mathematicians as well find (2.6) more convenient than the unwieldly (2.4).

The digression into Fourier analysis is important because it assures us that everything we have discovered about digitizing sine waves can be generalized to digitizing any period function. In particular, by treating the speech signal as if it is made up of periodic functions, which for the most part it is, we can apply the principles of sampling and quantization to digitizing speech.

A graph of the speech signal in the time domain shows time on the x-axis and signal amplitude on the y-axis, as in Figure 2.2. To digitize this signal we must sample it. At what rate should we sample to ensure that vital information is not lost to aliasing? Fourier analysis tells us that the speech signal is comprised of sinusoidal waves of varying frequencies. To represent the speech signal faithfully we must choose a sampling rate greater than twice the highest frequency of any sinusoidal component (see Fact 3 in Section 2.3.1). Taken literally, this proves difficult. Speech has frequency components well beyond the range of normal hearing, about 20,000 Hz. If we were compulsive about not misrepresenting any information through aliasing, we would take 40,000 samples each second. With a 12 bit quantizer we would generate a megabit of data every two seconds. While today's hardware can handle such sampling, both in terms of speed and storage, it's not practical.

Most of the information in speech is contained in the first three formants, as discussed in Chapter 1. Formant frequencies up through the third formant rarely exceed 4,000 Hz. Although higher frequencies play a small role in speech perception, helping to distinguish fricatives such as [s] and [f], and providing information about the identity of the speaker, speech is comprehensible without them. This suggests that we can get away with a sampling rate as low as 8,000 samples per second.

What would happen if we sampled raw speech at 8,000 samples per second? We would faithfully represent at least the first three formants, but the signal would be noisy, perhaps unintelligible, when it was reconstructed. That is because aliasing would cause the higher frequencies to be reconstructed as lower frequencies (see Fact 4 in Section 2.3.1), and these would be heard as noise in the lower frequency ranges.

To avoid this problem we preprocess speech through a **filter**. A filter takes sound of any frequencies as input, and outputs sound only in certain frequency ranges. A **low-pass filter** with a cutoff of N Hz outputs only frequencies of N Hz or less. A **high-pass filter** with the same cutoff outputs only frequencies of N Hz or greater. There are also **band-pass filters** that output frequencies within a certain range, or ranges. These are illustrated in Figure 2.16.

Filters are not perfect. A low-pass filter with a cutoff of 4,000 Hz will allow some frequencies slightly above 4,000 to leak through. If we graphed the output of an ideal low-pass filter with cutoff N it would look like Figure 2.17. In reality, we get something more like Figure 2.18. The shaded area in Figure 2.18 indicates the presence of unwanted sound or noise. Filters, then, are another source of noise.

Commonly, when speech is processed it is first low-pass filtered with a cutoff of 4,000 Hz. This guarantees that a sampling rate of 8,000 or higher will not result in very much aliasing, depending on the effectiveness of the filter. In practice it's best to either low-pass filter somewhat below twice the intended sampling rate, or sample at a rate somewhat above twice the filter cutoff frequency.

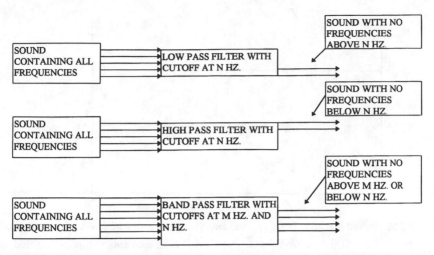

Figure 2.16 Filters attenuate certain frequencies while allowing other frequencies to pass through.

2.4.1 Wave Form Coders

The process of representing speech by sampling and quantizing the analog signal after filtering is called **wave form coding.** It results in a time dependent sequence of numbers that encodes the speech digitally. The inverse process of reconstruction, or **wave form decoding,** reproduces an approximation to the original analog signal. Figure 2.19 illustrates the process.

When nothing more is done to the numbers after sampling and quantization, the particular wave form coder is called **pulse code modulation** (PCM).

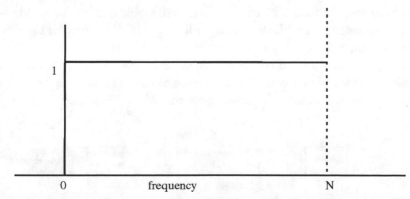

Figure 2.17 Output of an ideal low-pass filter with cutoff at N Hz. 0 = no output; 1 = 100% of original input.

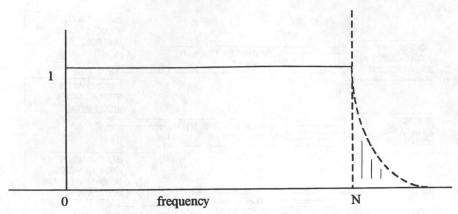

Figure 2.18 Output of a non-ideal (realistic) low-pass filter with cutoff at *N* Hz. 0 = no output; 1 = 100% of original input. The shaded area indicates that non-permitted frequencies have leaked through.

PCM is the simplest and most robust of the wave form coders. It is also the least efficient. A typical PCM coder—the one used by most telephone companies in North America—has a sampling rate of 8,000 samples per second with an 8 bit quantizer. Thus one second of telephone quality speech requires 64,000 bits of storage. Using PCM to get high quality music audio you need to sample 44,000 times per second (to cover frequencies up to 22,000 Hz, the limit of human hearing). Thus each channel of music recorded electronically requires 352,000 bits per second. A typical symphony by Anton Bruckner, not noted for brevity, will gobble up around 6 gigabits in quadraphonic, barely leaving room for Bolero on the compact disc.

PCM is like eating at a McDonalds restaurant. No matter where in the world you are, the food is of a known and predictable quality. No matter what input is encoded by PCM, the quality will be predictable, affected only by filtering and quantization limitations.

The fundamental theorem of digital data transmission is this: bits is money. Most readers have had the experience of waiting for a long file to download, perhaps in the form of a picture from the World Wide Web. If you waited

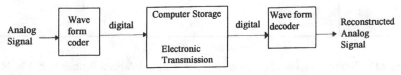

Figure 2.19 Wave form coding and decoding.

more than a minute, as well you might during peak Internet hours, and you value your time at $15/hour, then your one minute wait cost you a quarter. If you could pay ten cents and double the speed of transmission, you'd be ahead of the game.

How might you increase the speed of transmission? One way is to push through more bits per second, perhaps by adding more capacity at some bottleneck. Another way is to reduce the number of bits being sent, but with little or no information loss. This is a process called **compression**. Compression is ubiquitous throughout the computer and telecommunications industries. Files are "zipped" into smaller, more manageable files for storage and transmission. When they are needed they are "unzipped." "Zip" is just computer jargon for "compress."

The basic technique for compressing digitized speech is **difference coding.** Here's how it works. Suppose you had a sequence of numbers whose values change slowly: 942, 942, 941, 940, 941, 943, 944, etc. As written, the sequence of seven numbers is represented by 21 characters (ignoring spaces and commas). Suppose I write them this way: 942, 0, −1, −1, +1, +2, +1. This is an encoding of the first sequence. To decode you take the first value literally. For the second value you add or subtract the second number to the previous value. Continue this process, always adding or subtracting a small value to the previously decoded value. Using this form of difference coding, the sequence of 7 numbers can be represented by 14 characters.

This is the idea that underlies a second type of wave form coder called **delta PCM.** When sampling begins, the first value is the amplitude of the first sampled point. The second value is the difference between the amplitudes of the first and second sampled points. However, if the difference exceeds the range of the quantizer (a 4-bit quantizer can only record values between −8 and +7, for example) then it's set to the value of greatest magnitude allowed by the quantizer. The next value is the difference between the current amplitude and the previous amplitude as re-computed from the previous quantized difference, and so on. Computationally, the process is simple, requiring for each point only an addition to reconstruct the previous amplitude, and a subtraction of that value from the current sample.

Figure 2.20 illustrates the encoding process.

Table 2.2 illustrates the encoding. Suppose we have a 4-bit quantizer with its 16 quantization levels at integral values between −7 and +8. The first sampled value, 900, is stored and becomes the previous value. The next two samples are within range of the quantizer, so the differences are stored exactly, and the previous recomputed sampled value is the same as the previous sampled value. In the fourth row, however, the difference is 10, which is stored as 8. The previous recomputed sample value is 911—2 less than the actual value. It is the

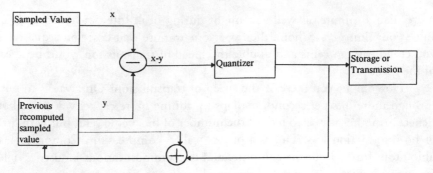

Figure 2.20 Delta PCM Encoder. It takes the current sampled value (x) and the recomputed previous sampled value (y) as its input. In Step 1, it outputs their difference (x − y). In Step 2 the difference is subject to quantization, which will only affect its value if it exceeds the range of the quantizer. In Step 3 the quantized difference is added to the current previous recomputed value to give the next recomputed value. The process repeats these three steps.

basis for the next difference which is 12. Again, the best the encoder can do is 8. The encoder is falling farther behind. This continues until the changes between sampled values becomes less. Now the encoder can catch up. Even though the difference between the sampled values 946 and 943 is only 3, the encoder stores a value of 8, allowing the previous recomputed sampled value to gain ground. The final difference shown is only 1, but the encoder catches up entirely by recording a 4.

Table 2.2
An Illustration of the Delta PCM Encoding Process

Sampled Value	Previous Recomputed Sampled Value	Computation Giving Previous Column	Value Stored by Delta PCM Encoder	Computation Giving Previous Column
900	900	initial value	900	initial value
901	901	900 + 1	1	901 − 900
903	903	901 + 2	2	903 − 901
913	911	903 + 8	8	913 − 903 → 8
923	919	911 + 8	8	923 − 911 → 8
933	927	919 + 8	8	933 − 919 → 8
943	935	927 + 8	8	943 − 927 → 8
946	943	935 + 8	8	946 − 935 → 8
947	947	943 + 4	4	947 − 943

The reconstruction process is simpler, still. It requires the addition of a difference value to a previously computed value, which becomes the next previous value, as suggested in Figure 2.21. The column (2) of previous recomputed sampled values is precisely what the decoder of Figure 2.21 will reconstruct.

Provided that the changes are for the most part small—say mostly between -8 and $+7$—four bits per sample suffices to represent the signal, plus some trivial overhead, at a fifty percent savings over 8-bit PCM. Since speech is for the most part a slowly varying signal, i.e., the differences are small, delta PCM works fairly well. If the signal changes too rapidly in places, however, quantizer noise will appear in the reconstruction, as would be the case in the illustration of Table 2.2.

When bits are at a premium and speech quality is not an issue, delta PCM can be implemented using one bit per sample. This corresponds to a quantizer of resolution 1, with two levels of quantization. If the next sampled value is larger then the previous, it is represented by a 1; if smaller, it is represented by a 0, which is interpreted as -1. This ultimate form of difference coding is called **delta modulation.** For a slowly varying signal, delta modulation provides an acceptably faithful representation. If the signal varies rapidly, however, a condition called **slope overload distortion** occurs in which the coder falls behind. In Figure 2.22 the step function of delta modulation approximates the continuous function it encodes. Slope overload distortion will lead to some error when the signal is reconstructed.

Each quantizer step in delta modulation must be a positive or negative value. It cannot be zero. During periods when the signal is unvarying, delta modulation forces small variations anyway. This error is called **granular noise.** It also occurs during periods of silence when the signal amplitude is zero for stretches of time. In that case the error is called **quieting noise** (see Figure 2.22).

To reduce the effect of slope overload distortion, granular noise and quieting noise, the signal may be sampled at rates considerably beyond the Nyquist rate. The effect of this is to make a rapidly changing signal more easily approximated in single steps since the amount of change between sampled points is

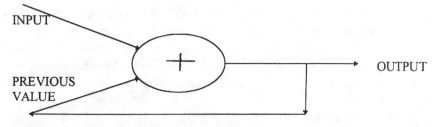

Figure 2.21 Delta PCM Decoder. It adds the current input (a difference) to the previously computed value, which then becomes the next previously computed value.

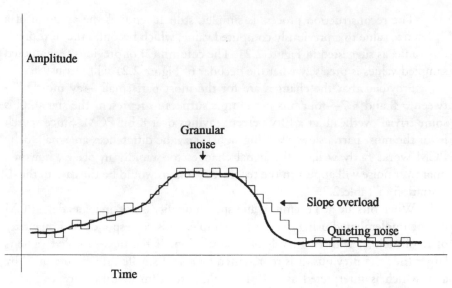

Figure 2.22 Delta modulation with slope overload, granular noise and quieting noise.

smaller. This reduces some of the advantage of one bit per sample. Instead of sampling 8,000 times per second, a coder using delta modulation might sample at 32,000 times per second. Thus delta PCM at four bits per sample and a sampling rate of 8,000 has about the same storage requirements as delta modulation with a sampling rate of 32,000.

We can build a little intelligence into the delta modulation process giving us **continuously variable slope delta modulation (CVSD)**. When an unvarying sequence of all 1's or all 0's occurs, the presence of slope overload may be inferred. The quantizer step size is then increased. Now a 0 or 1 jumps two steps instead of one. This improves the approximation. If the unchanging sequence continues, the step size may be increased even more. This is illustrated in Figure 2.23.

The quantizer step size increase is allowed to fade out. After a mixed sequence of 1's and 0's, the step size is reduced. This is also shown in Figure 2.23. The parameter ten is arbitrary and may be adjusted to suit the application.

In implementing CVSD it is important that both encoder and decoder are running the same algorithms for changing the quantizer step size. Otherwise they could get out of synch and the reconstructed signal would not represent the original one.

CVSD is an **adaptive coder**. The quantizer step size associated with the 0 or 1 output adapts to the signal being sampled. A more sophisticated adaptive coder is **adaptive delta PCM (ADPCM)**. It is based on delta PCM but uses a

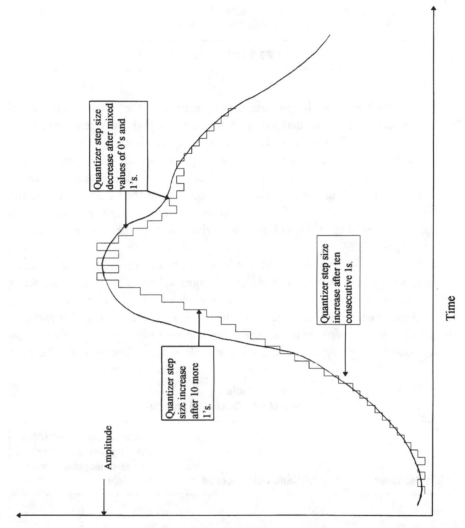

Figure 2.23 Continuously variable slope delta modulation (CVSD).

table lookup procedure to adjust the step size of the quantizer. In principle, it works this way. Suppose we have a quantizer step size table (Table 2.3) with the following values, where the highlighted numbers are the ones the coder is allowed to use at any given time. (In practice, the table would be much larger.)

Consider a 4-bit ADPCM coder. Each encoded value is a difference (except the first). Let one bit represent the sign of the difference, 1 for plus, 0 for minus. Let the other three bits be an index into the table of quantizer step sizes. The three bits can select from among some set of eight values somewhere in the

Table 2.3
Quantizer Step Size Table

0 1 2 3 5 7 9 11 14 17 20 24

table—shown highlighted—and use the one nearest to the actual magnitude of the difference as the encoded value. A predictor algorithm determines from which eight entries in the table the quantizer step size is chosen.

As illustration, consider Table 2.4. Suppose we are sampling a signal whose amplitudes are given in the first column of the table, and we are using delta PCM with a four bit quantizer. We would store the values shown in the second column of the table. The decoder would reconstruct the values shown in the first entry of the third column, with the resultant error shown next to it. The total error is 69.

Now consider how ADPCM would work on the same signal. At the start, table selection defaults to the highlighted eight values. The first difference is +5. That corresponds to an encoded value of +4 (5 is the fourth entry in the highlighted portion of the table counting from zero.) Assuming that the predictor algorithm acts to center as best it can the next highlighted portion of the table around the previous value used, the table remains unchanged. The next

Table 2.4
Example of 4-bit Delta PCM Encoding

Sampled Value	4-bit Delta PCM Encoded	Reconstructed Value (Also Previous Value for Encoder) The Error magnitude is on the Right.	
900	900	900	0
905	5	905	0
915	8	913	2
933	8	921	12
933	8	929	4
934	5	934	0
937	3	937	0
924	−7	930	6
914	−7	923	9
900	−7	916	16
889	−7	909	20

difference is 10 (915 − 905). The encoded value will be +6, which selects the seventh entry of the table, namely 9, to best represent the actual difference magnitude of 10. The highlighted (usable) part of the table is re-centered around 9 for the next sample: 0 1 2 **3 5 7 9 11 14 17 20** 24. The next sample is 933. The difference between it and 914, the previous reconstructed value, is 21. The eighth entry of the boldfaced part of the table, namely 20, is the closest difference. It is encoded as +7. Table 2.5 shows the entire encoded and reconstructed signal.

Table 2.5
Example of 4-bit ADPCM Encoding

Sampled Value	4-bit ADPCM Encoded (Table selection in parentheses)	Predictor Table	Reconstructed Value (also previous value for encoder) The error magnitude is on the right.	
900	900		900	0
		0 1 2 3 5 7 9 11 14 17 20 24		
905	+4 (5)		905	0
		0 1 2 **3 5 7 9 11** 14 17 20 24		
915	+6 (9)		914	1
		0 1 2 **3 5 7 9 11 14 17 20** 24		
933	+7 (20)		934	1
		0 1 2 **3 5 7 9 11 14 17 20** 24		
933	−0 (5)		929	4
		0 1 2 3 5 7 9 11 14 17 20 24		
934	+4 (5)		934	0
		0 1 **2 3 5 7 9 11 14** 17 20 24		
937	+3 (3)		937	0
		0 1 2 3 5 7 9 11 14 17 20 24		
924	−7 (11)		926	2
		0 1 2 **3 5 7 9 11 14 17 20 24**		
914	−3 (11)		915	1
		0 1 2 3 **5 7 9 11 14 17 20 24**		
900	−4 (14)		901	1
		0 1 2 **3 5 7 9 11 14** 17 20 24		
889	−3 (11)		890	1
		0 1 2 3 **5 7 9 11 14 17 20 24**		

The advantage of ADPCM is that it can better encode a signal that has both small and large variations by having many table entries of quantizer step sizes, some fine and some coarse, permitting the encoder to adapt to various circumstances. The total error in this illustration is 11, compared with delta PCM's 69, at the cost of the overhead of the more complex ADPCM software. The quantizer step size table and the predictor algorithms are present at both the encoding and decoding sites, so only a 4-bit value need be transmitted for each sample. The illustration given here is an oversimplification of the actual procedure but captures the fundamental idea. (See [3], pp. 440–444, for details.)

In practice, 4-bit ADPCM sampled at a rate of 8,000 is about as good as 8-bit PCM sampled at the same rate and is superior to CVSD sampled at 32,000. (CVSD's simplicity, once its selling point, has been obviated by modern hardware and software capabilities.) The *Consultative Committee for International Telegraph and Telephone* (CCITT), which sets international standards for the exchange of encoded speech, has established ADPCM as the standard coding method. The *American National Standards Institute* (ANSI) has done the same for telephone systems in North America.

Many telephone channels are capable of transmitting 64,000 bits per second (bps). This allows 4-bit ADPCM sampled at 16,000 bps to be transmitted. The higher sampling rate results in a reconstructed signal of excellent quality. It is preferred by the international community because of the importance of clarity in understanding non-native speakers of languages. It is also suitable for telephone conference calls because the speech is so accurately encoded that participants can often be identified by their voice alone and needn't preface remarks with their name every time they speak.

The final type of wave form coding we will consider is **subband coding** (**SBC**). Band-pass filters are used to partition the signal into frequency subbands. Each subband is encoded separately, usually by PCM or ADPCM. The advantage of subband coding is that lower bit rates can be used in less critical portions of the spectrum.

For example, most of the spectral energy in voiced speech is concentrated at the lower frequencies. By using 4-bit quantizers for the lower frequency bands and 2-bit quantizers for the higher frequency bands, some bit rate compression can be achieved with only a small loss in signal intelligibility after wave form decoding.

In this section we discussed four methods of waveform encoding: PCM, CVSD, ADPCM and SBC. In general PCM is the highest quality at a cost of 64,000 bps; ADPCM is of about the same quality at 32,000 bps. CVSD is of somewhat lower quality at 32,000 bps, and SBC of the lowest quality at 24,000 bps.

All of these basic methods have variants. Different algorithms for changing quantizer step sizes give us CVSD encoders of differing characteristics. The table size and arrangement for ADPCM affects how that encoding method will behave. The number of subbands and their coding rates leads to a large number of slightly different subband encoders.

2.4.2 Voice Coders (Vocoders)

A vocoder is an electronic model of the vocal tract capable of producing speech when provided with the proper input. It consists of an excitation component that generates periodic waveforms that simulate voiced speech and aperiodic waveforms, or white noise, that simulate unvoiced speech and other fricational components of speech. In addition, it contains banks of filters that approximate the effect of the various vowels and consonants. The filter parameters are computed from an analysis of the waveform of the speech signal that the vocoder is intended to replicate.

To use a vocoder, the original waveform is digitized and subjected to complex analyses to extract the appropriate filter parameters for the vocoder, whose output will then approximate the acoustical properties of the original signal. Since the vocoder itself, rather than a reconstruction of the original waveform, is the source of the speech, the term **source coder** is sometimes used.

Figure 2.24 shows the process schematically.

The process of parameter extraction for a vocoder is computationally intense but manageable with today's fast hardware and efficient algorithms. The parameters can be stored or transmitted at relatively low bit rates, anywhere

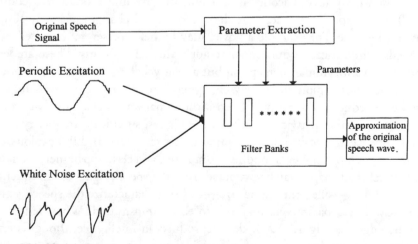

Figure 2.24 Model of a generic vocoder.

from 9,600 down to 1,200 bits per second of speech signal. (Compare with ADPCM at 32,000 bps of speech.) Once the parameters are computed, the original signal can be discarded. The spectral characteristics of the voice are quickly and simply reconstructed from the parameters by the vocoder, which can be installed wherever the speech signal itself is needed.

Vocoders are one of the oldest forms of speech processing. The first ones were put into use in the late 1930s. How did they do it without computers? How did they do anything without computers? (Clever devils, our ancestors.) The oldest kind of vocoder is called a channel vocoder. Its original inception was a complex arrangement of analog filters, rectifiers, A-to-D and D-to-A converters, pulse generators, noise generators, pitch detectors, voicing detectors, and lots of wire. Today, channel vocoders can be implemented using digital signal processing techniques on modern computers. They yield intelligible, if mechanical sounding, speech using 2,400 bps, 10 times more efficient than subband coding.

Other types of vocoders that have been used are formant vocoders, phase vocoders, and homomorphic or cepstral vocoders. The first two are similar to the channel vocoder with special emphasis on formant reconstruction in the one case, and on phase changes at the output of each filter in the other. Homomorphic vocoders are based on computing parameters from the inverse Fourier transform of the logarithm of the power spectrum of the original signal. (The Fourier transform and its inverse will be discussed in Section 2.5.) Details of all three of these vocoders, none of which are in wide use, may be found in [1] and [3].

2.4.2.1 Linear Predictive Coding

The most widely used vocoder is implemented by **linear predictive coding (LPC)**. LPC exploits the fact that the speech signal changes relatively slowly most of the time. For example a vowel may be held for several hundred milliseconds, while fricative durations are approximately 100 ms. These are very short intervals for human perception but a relatively long time for a computer.

We cannot exploit the slowly changing signal by reducing the sampling rate, hence reducing the bit rate. Even if the signal was unchanging, we would still need to sample at twice the rate of the highest significant frequency component of speech, because of the sampling theorem (fact 3). LPC exploits the fact that a pitch parameter, a loudness parameter, and certain parameters related to the vocal tract shape may be extracted from the speech signal as infrequently as every 30 ms—called the **frame size**—and still characterize the speech signal accurately. These parameters may be stored or transmitted. When fed to an LPC decoder (see Figure 2.25), they reproduce an intelligible, though somewhat robotic sounding version of the original speech.

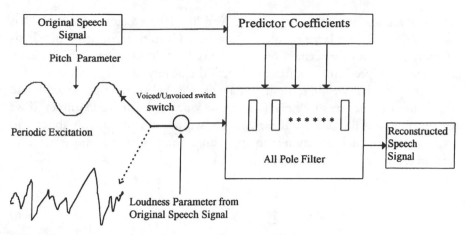

Figure 2.25 Model of an LPC decoder.

There are a myriad of implementations of the LPC method (see [3] for details). Frame sizes vary typically from 10 to 30 ms and may differ depending on whether the speech is voiced or unvoiced, as may the number of bits used to represent the parameters of a frame of speech. For voiced speech, it is common to use 10 parameters related to the vocal tract shape, plus two more for pitch and loudness. Unvoiced speech requires fewer vocal tract parameters, as little as four in some implementations. LPC is implemented at bit rates that generally range between 1,200 and 9,600 bps, the larger rates being needed if high intelligibility and more natural sounding speech is desired.

The vocal tract shape parameters are **predictor coefficients.** Here's how they are determined for the most commonly used LPC algorithm called LPC-10 because there are 10 predictor coefficients. Speech is sampled and quantized, and equations are set up under the assumption that any individual sample value is approximated by a linear combination of the previous 10 samples. Mathematically

$$S(n) \approx a(1)S(n-1) + a(2)S(n-2) + \ldots + a(10)\ S(n-10),\ n > 10$$

$$(2.7)$$

or more succinctly

$$S(n) \approx \sum_{i=1}^{10} a(i)S(n-i),\ n > 10$$

The $a(i)$ are the predictor coefficients. We'd like 10 predictor coefficients that work fairly well for a large number of samples. Then, to reconstruct hundreds of samples worth of speech, all we would need would be a few starting values and one set of predictor coefficients. That's the beauty of LPC.

Imagine that we are sampling at a rate of 8,000 samples per second, and we choose a frame size of 20 ms. Each frame would contain 160 samples. If we set up equations like 2.7 for these samples, we would end up with about 150 such equations. Each equation is an approximation, that is, it is only exact if we add an error term, $E(n)$, as shown in (2.9).

$$S(n) = \sum_{i=1}^{10} a(i)S(n - i) + E(n) \qquad (2.8)$$

The trick now is to find one optimal set of 10 predictor coefficients such that $E(n)^2$ is minimized over all available samples in the window. (Minimizing the square of the error takes care of both positive and negative errors.) This is mathematically feasible. Details may be found in many sources, for example [4].

What do these coefficients have to do with modeling the vocal tract? It turns out that when (2.9) is written in a certain mathematical form called z-transform notation, it becomes a mathematical model of a type of vocal tract filter known as an "all-pole digital filter." Such a filter focuses on spectral maxima so it is particularly sensitive to formants. The role of the predictor coefficients in the model is to control the filter in such a way as to replicate the formants and other frequency variations in the original speech. As Figure 2.25 suggests, the decoding process first makes a decision as to whether the speech is voiced or unvoiced, and provides periodic or aperiodic sound accordingly. The pitch parameter controls the frequency of the periodic wave generator. The loudness or gain parameter determines the intensity of the output. Finally, the "guts" of the decoder is the all-pole digital filter controlled by the predictor coefficients. It is computationally fast. Most of the work in compressing speech by LPC is computing the predictor coefficients on the encoding side of the process.

LPC does a good job replicating vowel sounds, because they depend most heavily on the location and intensity of formants. Consonants fare worse under LPC because the higher frequencies are not as well replicated as the lower ones due to the reliance on the all pole digital filter. Moreover, LPC is also weak in coding speech in the presence of noise. Small amounts of noise create large discrepancies in the predictor coefficients, with concomitant inaccuracies during decoding. The chief advantage of LPC is its low bit rate and good intelligibility in the absence of noise.

Refinements to the basic LPC process are possible. A **residual excited linear prediction (RELP) vocoder** computes and transmits a residual error. After computing the predictor coefficients using an ordinary LPC encoder, the speech is decoded before transmission and subtracted from the original speech signal to yield a residual error. Essentially, the error would be determined statistically from the various $E(n)$ of (2.9). This error is transmitted along with the predictor coefficients. At the receiving end, the speech is recreated in the usual way, and then the residual error added in to improve the accuracy of the wave form. At 9,600 bps of speech, it produces intelligible output suitable for telephone transmission.

Multipulse LPC vocoders employ a technique for raising the quality of LPC by improving the regeneration of higher frequencies (the ones needed to distinguish among [s], [f], and [th], for example). It produces highly intelligible speech at 9,600 bps. **Code-excited linear prediction (CELP) vocoders and vector sum excited linear predication (VSELP) vocoders** are both designed to preserve the quality of LPC speech while lowering the bit rate. CELP vocoders produce highly intelligible speech at 4,800 bps. VSELP vocoders at 8,000 bps are good enough to be accepted as the standard speech coder for use in digital cellular telephone systems in North America. Details regarding these highly sophisticated implementations of LPC are to be found in [3].

Vocoders are designed to replicate the speech of a single individual based on parameters extracted from that person's speech. If the speech occurs in a noisy environment, or more than one person is speaking, the parameters that are extracted are affected and the recreated speech may be garbled. Waveform coders, on the other hand, are relatively immune to noise, simply cloning it along with the speech.

No matter what the method of speech digitization, there is a tug-of-war between the need for compression and the need for quality after decoding. Because speech is redundant—that's why you can have a conversation in a noisy room—imperfect reconstructions may still be highly intelligible. The need for intelligibility varies with the application. For inexpensive children's toys, low intelligibility is often acceptable. For telephone transmission of human speech, higher intelligibility is needed.

2.5 The Frequency Domain

Speech can be regarded as consisting of a mixture of sounds of various frequencies, each frequency having a certain amplitude at a certain time. In this sense, speech is three-dimensional with coordinates of time, frequency, and amplitude. A display of speech in this coordinate system would look like Figure 2.26.

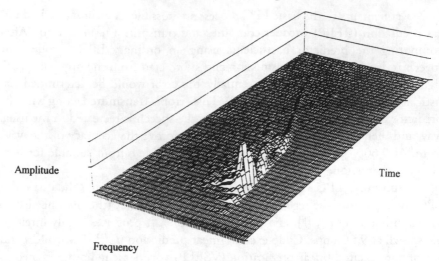

Amplitude

Time

Frequency

Figure 2.26 Three-dimensional display of speech.

Unfortunately, three dimensions are difficult to comprehend, especially when drawn on a two-dimensional surface such as a page in a book. In practice, we often eliminate the frequency dimension by combining the amplitudes of all frequencies at each point in time, yielding a single amplitude point, which is plotted against time. The result is a two-dimensional display like Figure 2.2 that, while concealing certain information, is easier to understand. This is a **time domain display** and is often the way a speech signal is represented since it captures the important fact that the signal varies in time.

The chief alternate way of representing the speech signal is a **frequency domain display.** In this case we look at a small interval of time and plot amplitude versus frequency. The resulting display is called a **spectrum.** Figure 2.27 shows a spectrum taken in about the middle of the pronunciation of [ē].

The amplitude of all frequencies between 0 and 5,000 Hz is shown. The curve is smooth, because values for missing frequencies are interpolated, and the amplitudes are averaged over the time interval. It is impossible in principle to know the distribution of frequency amplitudes at a point in time.[3] In Figure 2.27

---

3. This is a manifestation of a classical tradeoff in physics known as the Heisenberg Uncertainty Principle. Many readers will know it in the form: One cannot determine both the position and velocity of an elementary particle with complete accuracy; the more highly determined the one, the less highly determined the other. It is a consequence of the wave description of matter, and in the particular case of digital signal processing, of the wave description of sound. Intuitively, we cannot determine the frequencies in a very small time window with accuracy, because we have to observe the signal over a stretch of time to detect the periodicity.

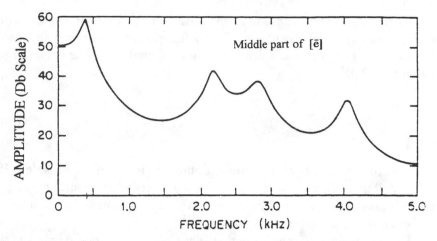

Figure 2.27 Spectrum of part of the pronunciation of [ē].

peaks of amplitude are visible at about 400, 2,100, and 2,800 Hz. These correspond to the first three formants. Their pattern is typical of the pronunciation of the high front vowel [ē] (the vowel in *beet*), as we saw in Chapter 1. The ordinate is measured in decibels, so it is a logarithmic scale. This has the effect of making the formant peaks stand out less than if the scale were linear. For example, the second formant peak at 2,100 Hz is seven or so decibels more than the dip at 2,500 Hz, or more than twice the amplitude. Since energy is proportional to the square of the amplitude, the second formant has over four times the amount of energy as the surrounding frequencies. As noted in Chapter 1, the disproportionate distribution of energy in the spectrum is perceived as vowel sounds.

How do we compute a spectrum? The answer lies in our earlier observation that speech can be represented in the time domain as a Fourier series, that is, as sums and differences of sinusoidals of varying periods, amplitudes, and phases. (This assumes that speech is periodic over short time intervals—not 100% accurate but a close enough approximation for practical purposes.) The sinusoidals represent the various frequency components of the signal. In this decomposed state, it is possible to compute amplitude as a function of frequency for a given interval of time, which is precisely what a spectrum is.

The computation relies on the notion of a "transform." In mathematics, a function may be transformed by defining a different function whose values are determined by the original function. There are several reasons for doing such a thing. Operations on the transformed function may be more easily carried out than on the original function. A second reason is that the transformed function may reveal physical properties that are not explicitly described by the original function.

There are many transforms in mathematics. One that is used frequently in speech processing is the z-transform. Given a function $x(t)$—whose values you may imagine to be samples of a signal starting at t (time) $= 0$—the z-transform is defined in (2.9).

$$X(z) = \sum_{t=0}^{\infty} x(t)z^{-t} \quad \text{or} \quad X(z) = \sum_{t=0}^{\infty} \frac{x(t)}{z^t} \qquad (2.9)$$

A function of the variable t is reformulated into a function of the variable z. If you know all the $x(t)$, you can compute $X(z)$ for values of z for which the summation $X(z)$ is finite. The independent variable z is complex, that is, it may have both a real and an imaginary part.[4]

There is an inverse transformation. Thus given $X(z)$, one recovers $x(t)$ from (2.10) (j indicates a complex variable):

$$x(t) = \frac{1}{2\pi j} \oint_C X(z)\, z^{n-1}\, dz \qquad (2.10)$$

The symbol $\oint_C$ in (2.10) is a contour integral where C is the contour within which $X(z)$ is defined. The significance of the inverse transformation has to do with the first reason for doing transforms: One can perform certain operations on the transformed function that might be difficult to perform on the original function. Once performed, the inverse transform is applied and the results carried back to the original function. Logarithms work in a similar way. One can "transform" two positive numbers into logarithms, add the logarithms in the "logarithm domain," and apply the inverse (anti-log or exponentiation) operation to that sum, which results in the product of the two original numbers. Older readers may remember this as a precalculator-days' method of multiplying long numbers, where the logs and anti-logs were found by table look-up.

The second reason for transforming a function is to see it from a point of view that makes certain physical properties manifest. If the z-transform is applied

---

4. I am well aware that many readers of this book may have forgotten, or perhaps never learned, the mathematics in the preceding, and several of the following paragraphs. If you are comfortable with trigonometry but not with complex exponentials, know that all the complex exponential expressions we use can be expressed in terms of sines and cosines. However, the complex exponential representation is standard and used throughout the field of speech processing, both in theory and applications. Please be assured that the essential ideas of this section may be understood without deep mathematical knowledge, allowing you to pass lightly over material that reminds you of Aldous Huxley's famous—some would say infamous—aphorism, "I admit that mathematical science is a good thing. But excessive devotion to it is a bad thing."

to the time domain representation of a periodic function with samples $F(0)$, $F(1), \ldots F(N-1)$, taken at a sampling rate of S samples per second, the result, after considerable mathematical manipulation, expresses amplitude as a function of frequency. One form of such a result is called the **discrete Fourier transform (DFT)**, given as (2.11).

$$X(k) = \sum_{t=0}^{N-1} F(t)e^{\frac{-2\pi jkt}{N}} \qquad k = 1, 2, \ldots N-1 \qquad (2.11)$$

Here, k is an independent variable related to frequency. $X(k)$ is the amplitudes of frequencies $f_k = k(S/2N)$ Hz for the $N-1$ values of k indicated in (2.11). These are equally spaced frequencies between zero and $S/2$ Hz. Frequencies beyond $S/2$ Hz are not present (have zero amplitudes), assuming the signal is low-pass filtered at $S/2$ Hz to prevent aliasing. From these discrete points, however, a spectrum such as Figure 2.27 may be computed. The significance of the transform is that it reveals the individual frequency components of the speech signal and their relative amplitudes, which is not evident from the time domain representation.

The **inverse discrete Fourier transform (IDFT)**, analogous to the inverse z-transform, is given in (2.12), showing how, ideally, the original time domain representation may be recovered. Certain conditions must be met for both the DFT and IDFT to exist. These conditions are generally satisfied by the mathematical approximations of speech that are commonly used. See [3] for details.

$$F(t) = \frac{1}{N} \sum_{k=0}^{N-1} X(k)e^{\frac{2\pi jkt}{N}} \qquad t = 1, 2, \ldots \tau - 1 \qquad (2.12)$$

Much practical speech analysis takes place in the frequency domain. Speech recognition, speech synthesis, voice (speaker) recognition, language identification, lip synching—all of these are discussed elsewhere in this book—are based on spectral analyses of one kind or another. Software is readily available on nearly any platform for transforming a sampled, quantized time domain speech signal into the frequency domain. The *fast Fourier transform* (FFT) family of computer programs have been around since the 1960s and books have been written on this subject alone (see [5], for example).

2.5.1 The Game of Jumble: Spectrum-Cepstrum, Frequency-Quefrency, Filtering-Liftering

A time domain representation of speech may be transformed into a frequency domain representation by an application of the DFT, producing the speech spectrum. The spectrum $|S(f)|$ may be viewed as the product of a slowly varying part $\Theta(f)$ and a quickly varying part $E(f)$. The quickly varying part, or excitation, is the result of vocal cord vibration. The slowly varying part corresponds to the movement of the articulators as they form the speech into vowels and consonants. Mathematically, we have

$$|S(f)| = |\Theta(f)| \times |E(f)| \qquad (2.13)$$

If we wanted to see this effect in the time domain, we would have to apply the IDFT. However, the IDFT is a linear transformation. This means that the IDFT applied to $g + h$ has the same result as the IDFT applied to g, plus the IDFT applied to h. The IDFT should not be applied to a nonlinear combination of functions. In the case of (2.13), to make a linear expression, logarithms are taken of both sides, resulting in (2.14).

$$\log(|S(f)|) = \log(|\Theta(f)|) + \log(|E(f)|) \qquad (2.14)$$

The IDFT is applied to (2.14) transforming it back into the time domain in such a way that the values of the slowly varying part occur at lower positions on the time axis than the values of the quickly varying part. This special time domain representation has "spectrum-like" properties because it discriminates between quickly varying (high "frequency") and slowly varying (low "frequency") parts of the signal. Therefore the term *cepstrum*—an anagram of *spectrum*—is applied to it. The discrete values that are output by the IDFT are called *cepstral coefficients*. Values on the time axis are given the name *quefrency* because of their parallel to frequency in the spectrum. In effect, the cepstral analysis separates out the effect of the articulators—low quefrency—from the periodic excitation of the vocal cords, found in the high quefrency ranges.

Just as we can filter a spectrum to eliminate or isolate certain frequencies, we can *lifter* the cepstrum to isolate certain ranges of quefrency. For example a low-time lifter—analogous to a low-pass filter—will isolate cepstral values in the low quefrency ranges; the converse is true for a high-time lifter. Liftering may be applied in the quefrency domain, the DFT applied to transform the signal, once again, into the frequency domain, where the spectrum derived from the "liftered" signal may have useful properties.

The spectrum reveals certain important characteristics of speech such as the location of formants. The cepstrum also reveals speech properties that are effective in speech recognition and other application areas. For example, cepstral coefficients may be used to model speech utterances for computer speech recognition. Their strength lies in their ability to ignore certain aspects of speech such as pitch. This is important for speech recognition, where it is desirable to recognize human speech whether spoken by a large man with a deep voice, or a female child with a high-pitched voice. Speech recognition is discussed in Chapter 3.

Cepstral analysis is practical within a limited scope, but we should bear in mind that it is "based on a few tenuous assumptions that can cause unpredictable outcomes" ([3], p. 362).

The mathematical analysis of the properties of the speech signal is one of the most highly developed theoretical areas of computer and electrical engineering. Nonetheless, our ability to process speech in a computer is a mere shadow of the speech processing capabilities of a five-year-old child, who easily recognizes people by their voices, comprehends continuously spoken speech at noisy birthday parties, and can tell you instantly whether you're speaking English or a foreign language. The treatise in [3] is 850 pages, many of them filled with heart-stopping equations. Nonetheless, Einstein's words are apropos:

> As far as the laws of mathematics refer to reality, they are not certain, and
> as far as they are certain, they do not refer to reality.

2.5.2 Spectrograms: A Hybrid Representation of Speech

Both time and frequency domain representations of speech are unsatisfactory in that one of the three dimensions of sound is suppressed: frequency in the one case, and time in the other. In a spectrogram, frequency and amplitude are plotted against time. Frequency is represented by position on the y-axis; the amplitude at a given frequency is represented by the relative darkness of the display. This is illustrated in Figure 2.28, a repeat of Figure 1.5.

Before the advent of the digital computer, an electronic device called a sound spectrograph would produce a spectrogram. It took a time domain representation of the speech signal, ran it through a bank of filters to isolate frequency bands, and recorded the output of the filters on a paper strip chart as frequency versus time, with higher amplitudes producing a more intense marking.

Today, spectrograms are generated in digital computers by taking a sampled time domain representation and computing a series of spectra in successive intervals, or **windows,** of time. FFTs are generally used. If the window is short, 2–5 ms, the spectrogram has good time resolution, but the frequency resolution is poor. That's because there isn't a big enough stretch of time to see all of the periodicity that may be present. On the other hand, if the window is long, around 20 ms, the frequency resolution is good, but resolution in the time dimension is fuzzy. That's because you can only determine where events

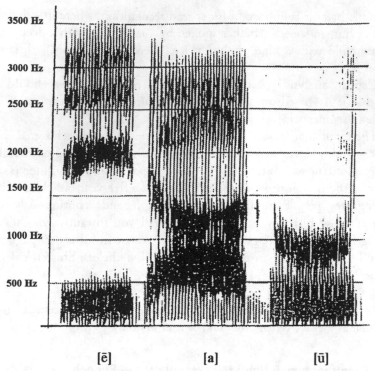

[ē]　　　　　　　　[a]　　　　　　　　[ū]

Figure 2.28 Spectrogram for the vowels [ē] as in *beet*, [a] as in *balm*, and [ū] as in *boot*.

begin or end within a tolerance of 20 ms, a relatively long stretch of time when you consider that a single vocal cord vibration has a duration of 5–10 ms. This trade-off, either good frequency resolution or good time resolution, but never both, is an unavoidable consequence of the physics of wave mechanics, as discussed in footnote 3.

With short windows and good time resolution, the resulting spectrogram is called **wideband** because the imprecision in the frequency domain causes blurring over a wide range of frequencies. A wideband spectrogram is illustrated in Figure 2.29.

The vertical striations each represent a glottal pulse, that is, a single vibration of the vocal cords. These are visible because of the excellent time resolution. Wideband spectrograms are useful for determining the length of phonetic events such as the period of silence preceding the release of a stop, or the duration of a vowel. The formants can be seen in this display as dark, more or less horizontal bands, but the individual harmonics that comprise the formant are blurred.

With long windows the resulting spectrogram is called **narrowband**, as illustrated in Figure 2.30.

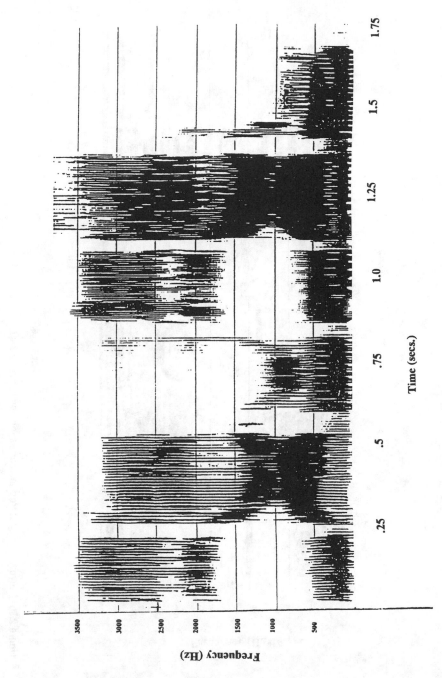

Figure 2.29 Wideband spectrogram of the syllables [dē], [da], and [dū] repeated twice, the second time at a higher pitch.

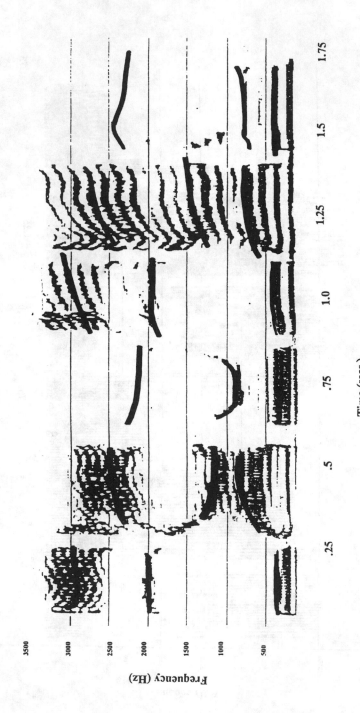

Figure 2.30 Narrowband spectrogram corresponding to Figure 2.29.

The speech that produced this spectrogram is the same is that of Figure 2.29. The smeared appearance of the formants in Figure 2.29 have resolved to the individual narrow bands of harmonics that constitute each formant. The vertical striations of the glottal pulse, however, are barely seen because time domain resolution is not fine enough to detect the relatively short single glottal pulse of six to eight milliseconds. Narrowband spectrograms are useful for analyzing the frequency distribution of speech. For example the syllable between 1.15 and 1.4 seconds shows 18 harmonics quite clearly. Since each harmonic is a multiple of the glottal pulse rate, that rate can be determined accurately by taking a relatively stable harmonic such as the third striation up from the bottom and dividing it by three. This gives us a glottal pulse rate of about 166 Hz. This can be verified in the wideband spectrogram by counting the glottal pulses in that same time span, 1.15 to 1.4 seconds. There are 41, more or less. Forty-one pulses in .25 seconds indicates a glottal pulse rate of 164 Hz.

As we saw in Chapter 1, each vowel has a characteristic formant structure. Consider Figures 2.29 and 2.30. The first 0.3 seconds show formants at around 300, 2,000, and 2,900 Hz, the characteristic pattern of a high front vowel such as [ē]. The short pause—the vertical white space—indicates a stop consonant. It is followed by a vowel whose formants are at about 900, 1,200, and 2,400 Hz, indicative of a low back vowel such as [a]. The third vowel has the low first formant typical of a high vowel and the relatively small spread between the first and second formants typical of a back vowel, suggesting a high back vowel such as [ū]. The utterance, in fact, consists of the syllables [dē], [da], and [dū]. The second three utterances were the same syllables pronounced at a higher pitch. The formants are proportionately higher but still retain the same relative proximities.

The art of reading spectrograms can be developed with practice and instruction. Indeed, you just had your first lesson: One identifies vowels by formant structure and stops by vertical blankness. With more instruction, and some practice, you can learn to distinguish voiced from unvoiced consonants and to recognize places of articulation such as bilabial, alveolar, and velar. You can learn to recognize the characteristic random noise pattern, especially in the higher frequencies, of fricatives and affricates. Nasals such as [m] and [n] have typical formant structures that are discernible, and [l]s, [r]s, [w]s and [y]s can also be detected by the careful eye [6].

Spectrograms have been called "voiceprints" in the popular culture and media. It has been suggested that each individual's voice is as unique to that person as their fingerprints. This may be true, but the spectrogram does not reveal the level of detail necessary to perform reliable speaker recognition, the kind that would hold up in a courtroom.

Still, the higher formants are more closely related to the physiological aspects of the vocal tract than to what is being said. These, of course, are individual traits. Other observable features that are speaker-, rather than language-dependent are average pitch, degree of aspiration, vowel length, relative vowel quality, speech rate, degree of vowel nasality, etc. We can distinguish women and children from men by pitch. A large man with a large head has a lower fourth formant than a small man with a small head. Indeed, all the formants of a large-headed person will be lower than those of a smaller headed person, other factors being the same. We can distinguish native speakers of English from non-native speakers in many cases. Native speakers of Romance languages who learn English almost inevitably fail to diphthongize vowels as native American English speakers do, and this is detectable by the relatively horizontal formants in spectrograms of their speech. All these factors might help us distinguish individuals by their spectrograms alone, especially if they are saying the same thing.

In Chapter 5 the subject of speaker recognition is considered. There we will see that it is possible to identify individuals by their voices alone with a high degree of accuracy. However, spectrograms will not play a role in modern detection methods.

References

[1]   Rabiner, L. R., and R. W. Schafer, *Digital Processing of Speech Signals*, Englewood Cliffs, New Jersey: Prentice-Hall, Inc., 1978.

[2]   James, J. F., *A Student's Guide to Fourier Transforms*, Cambridge, U.K.: University of Cambridge Press, 1995.

[3]   Deller, J. R. Jr., J. G. Proakis, and J. H. L. Hansen, *Discrete-Time Processing of Speech Signals*, New York: Macmillan Publishing Company, 1993.

[4]   Papamichalis, P. E., *Practical Approaches to Speech Coding*, Englewood Cliffs, NJ: Prentice-Hall, Inc., 1987.

[5]   Blahut, R. E., *Fast Algorithms for Digital Signal Processing*, Reading, MA: Addison-Wesley Publishing Company, Inc., 1985.

[6]   Ladefoged, P., *A Course in Phonetics, 3rd Edition*, Fort Worth, TX: Harcourt Brace Jovanovich College Publishers, 1985.

3

Speech Recognition

The first generations of computers had received their inputs through glorified typewriter keyboards, and had replied through high-speed printers and visual displays. HAL could do this when necessary, but most of his communication with his shipmates was by means of the spoken word.

Arthur C. Clark, *2001: A Space Odyssey*

3.1 Introduction

The notion of speaking to our nonhuman workmates is ages old. Draft animals were voice-trained before history was written. A good sheep dog understands sundry voice commands. The industrial revolution changed the way we communicate with our assistants. Wheels, pedals, and levers replaced the spoken word. Nonetheless, on a cold morning when Lizzie won't crank, desperate men and women resort to speech, and implore their inert metal beasts to stir.

Just when we've gotten used to the idea that talking to machines is futile, it isn't. Telephones may now be "dialed" by speaking the name of the person you wish to call. Many functions in the home can be voice-activated, and the brooding, silent monsters that once filled entire computer rooms have evolved into jaunty little notebooks that will understand your speech and talk back if you so desire.

If anything, the pendulum has swung the other way. To our parents and grandparents, the idea of talking to a machine and succeeding in doing anything other than venting frustration was incomprehensible. Our children and grandchildren, and perhaps we ourselves, possess exaggerated expectations when it comes to talking to computers, begotten by television programs such as *Knight Rider,* centering on a car-computer that converses fluently, and by the urbane HAL of motion picture fame[1], a space-traveling computer whose ability to recognize speech includes lip reading.

In this chapter, we will explore computer speech recognition from several points of view. First, we will define it. Then we'll trace its history. Next we will examine types of speech recognition, and then we'll get into the nitty-gritty of how it's done. Finally we'll look at the robustness of speech recognition and what the future holds.

3.2 Speech Recognition: What It Is; What It Isn't

It isn't understanding. Some years ago I wrote a chapter in a book in which I said, "The more I try to understand what it means for a machine to understand, the less I understand" [1]. I am stuck in this lamentable state despite the efforts of cognitive scientists. Language comprehension by a machine is one of the areas of concern to *artificial intelligence* (AI) experts. Their opinions run the gamut from "yes, it's possible and it's already happening" to "no, it's impossible." The question of whether a computer can be conscious enters into the equation, with respected, well-reputed scholars arguing all sides of the issues.[2]

Certainly there are degrees of understanding, putting the question of consciousness aside. It is useful to note the two extremes. I don't believe anyone would say that an automobile understands that it's supposed to stop when the brake is applied except, perhaps, metaphorically. At the opposite end, a computer capable of passing the Turing test, in the view of most—but not all—scientists, would be said to understand language. The Turing test, in a form somewhat modified from Turing's original description in [3], is conducted as follows: Behind two screens are a computer and a human being. An interrogator attempts to decide which is which (who is who?) by asking questions and evaluating the answers. If the interrogator is unable to do so in a decisive manner, the computer is said to have passed the test. To date, no computer has come close to passing the Turing test. There is much argument in the academic world about the legitimacy and even the possibility of such a test.

---

1. *2001: A Space Odyssey,* directed by Stanley Kubrick, written by Arthur C. Clarke.
2. Roger Penrose, Marvin Minsky, John Searle, Joseph Weizenbaum, Douglas Hofstadter, Hubert Dreyfus, Doug Lenant, and Edward Feigenbaum are a smattering of scholars who think and write on these matters. A discussion of the issues and some of their views may be found in [2].

Between the two extremes are computers that take as input complex commands in English (and other languages) and respond in complex ways. For example, in a context of data about naval ships, a computer could answer spoken questions such as, "What's the Mercury's average cruising speed?" or "What is the name and c-code of the carrier in Siberian Sea?" [4]. The computer answers correctly—but did it understand the questions?

What I do know is that the computer must recognize most of the words in the question in the sense of being able to repeat them back correctly, much as a shorthand secretary can after taking dictation. This suggests an answer to the "What it is" part of this section's heading. **Computer speech recognition** is the ability to take speech as input and produce a transcript of that speech as output.

Computer speech understanding, to the extent it exists at all, requires speech recognition as a precursor, or perhaps more accurately, as a "cocursor"—understanding and recognition go hand-in-hand. Conversely, speech recognition, to be accurate, requires computational processes that are part of speech understanding, such as a syntactic and semantic analysis of the speech.

3.3 Why Is Speech Recognition Easy for Us and Difficult for Our Computers?

The human system of hearing is capable of complex analysis. Through ingenious and highly evolved mechanisms, the ear performs spectral decomposition of auditory input and conveys the information to the brain where it is interpreted. Sounds as diverse as gunshots, wind rustling in the leaves, telephones ringing, and the allophones of speech are all easily recognized in context.

Context is essential. Alone at night in a strange house, the brain may interpret benign sounds as ominous. An acorn rolling across the roof sounds like footsteps in the attic; a loose shutter in the wind is "The Stalker" forcing a window. Language is no different. In isolation, the vowel sound [o] would probably not be heard as a vowel sound. Put it in the context [k_t] ("c_t") and it will be heard as a vowel sound, though without further context it may be confused with the vowel sound [ô], as in the vowel in the word *caught*. Expand the context to "army [k_t]" and most people will hear the [o] sound of *cot*. The same vowel sound in the context of "[k_t] the ball" will likely be heard as [ô].

Language is the context in which speech sounds are recognized. Language includes words, phrases, sentences, their meaning, and the larger context required to comprehend meaning. This "larger context" includes previous utterances, immediate surroundings, and general knowledge of the world. The sentence *The book was* [red] cannot be interpreted from the speech sounds alone.

It may be "The book was read (by the teacher)", or "The book was red (in color)." Context is needed for a final decision. Similarly, the phrase *synthetic buffalo hides,* which allows for two interpretations—"hides made from synthetic materials," or "hides made from synthetic buffaloes"—is correctly interpreted because our general knowledge informs us that synthetic buffaloes are not yet possible. (However, sheep were cloned during the writing of this book. Are they synthetic? If so, then synthetic buffalo cannot be far behind, and my illustration is dated.)

In Chapter 1 we saw that casually spoken speech may undergo articulatory effects, so a sentence such as *"Did you eat yet?"* may sound like *geechet.* People readily recognize abridged speech most of the time. It occurs with more frequency than you might think and poses serious problems for computer speech recognition. The computer is required to deduce from context the meaning of the question and then use the few phonetic clues available to reconstruct the underlying sentence. While one can imagine this to be possible on a case-by-case basis, designing algorithms to reconstruct abridged speech in a general way has yet to be accomplished.

In rapid speech, the aspiration of the /p/ in *parking* may not be present, making the word sound exactly like *barking.* In principle, neither a person nor a computer could determine which word the speaker intended to say from the sounds alone. This situation occurs repeatedly in speech: /s/ and /f/ sound similar on the telephone; /i/ and /e/ sound the same when they occur before a nasal consonant in many American dialects. When you recognize speech you are performing rapid-fire logic based on context, your knowledge of what words mean, and your knowledge about the world. Take this test:

a) As far as poodles are concerned there are fewer problems with _arking. (/p/ or /b/)

b) If the shoe _its, wear it. (/s/ or /f/)

c) You'll need a p_n to take this quiz. (/i/ or /e/)

Consider the inferences. You have to know a poodle is a kind of dog and that barking is more frequently associated with dogs than parking. However, it's not necessary logic. I can make it go the other way by adjusting the context: All the dog groomers work downtown except the one that does poodles. So as far as poodles are concerned, there are fewer problems with _arking. Now the inferences are different. You have to know that it is generally difficult to park downtown, and draw your conclusions based on that fact. If you're bored, think up similar contexts to force (b) and (c) into one or another possible interpretation.

These almost trivial mental exercises (for humans) are difficult for computers. It's not the logic. Computers excel at logic. They are the electronic embodiment of logic. The problem is the prodigiously intricate webs of meaning and real world associations that every human being carries in his head and accesses effortlessly when recognizing language.

When it comes to pure spectral analysis, computers can exceed human ability. A high enough sampling rate will detect frequencies above the range of human hearing. Couple this with a large quantizer, nuances of sound beyond the capacity of humans to detect are possible. Even the ear's remarkable ability to process sounds differing vastly in loudness can be exceeded by modern electronics.

Computers can store huge dictionaries containing more words and their definitions than any human being could ever know in a lifetime. Computers can parse, that is, syntactically decompose ("diagram") sentences in an instant. However, instilling meaning and contextual awareness in a computer in the deep sense that it is found in the human brain eludes computational linguists.

That is why we haven't yet replaced court stenographers with computers. That is why you type when you might speak. That is why billions (with a *b*) of dollars are spent in the medical industry transcribing by ear and hand the spoken notations of physicians, though in one special instance, that of radiology reports, computer speech recognition is being deployed with success.

Despite the challenges, there have been many partial successes with speech recognition. We'll take a look at some of them in Chapter 6. Nonetheless, a computer's ability to recognize continuously spoken speech in the presence of background noise, in a non-predetermined context—a feat within the capabilities of most human beings—is a goal currently beyond the pale, though we are striving mightily to get there.

3.4 A Brief History of Speech Recognition

Unlike speech synthesis, whose history goes back to the eighteenth century, speech recognition only became practical with the invention of modern electronics.

The earliest attempts at speech recognition were based on sound spectrographs. From their spectral display, vowels may be recognized, stops and fricatives detected, and the presence of voicing determined. This amount of information is adequate for recognizing a vocabulary consisting, say, of the first ten digits: zero through nine, as spoken by the same person.

Operationally, the person speaks each of the ten words into the microphone of the speech recognizer. Parameters are extracted from the spectral patterns of each word and stored in an electronic memory, where they are called **templates.** Each vocabulary word has its own template as spoken by a particular speaker. The process in which templates are created is called the **training stage** of speech recognition.

At some later time, a speaker may pronounce any one of the formerly trained words into the microphone of the recognizer. From this utterance a template of the word is created in precisely the same way as it was when the word was trained previously by that speaker. That template is compared with the templates previously created by that speaker. The best match determines the word that the system recognizes. Figure 3.1 illustrates the process.

If the speaker wanted to speak several words, each utterance had to be separated by a pause of about a half-second duration. This type of speech recognition is called **isolated-word recognition** or **discrete-word recognition**. Early recognizers were only capable of dealing with one word, or more precisely, with one "templated" utterance at a time. (One could make a single template out of several words if so desired, but the system would treat it as if it were a single "word.")

Under ideal conditions, such systems achieved fairly high accuracy, correctly identifying which digit was spoken over ninety-five percent of the time. The operative word here is "ideal," and it continues to this day to be central. Ideally, the same speaker who trained the system, and from whose pronunciations the templates were created, must test the system. A change in speaker resulted in a drastic reduction in performance.

Ideally, the speaker had to enunciate the digits distinctly, one at a time, and in the same manner as in the training phase. If he dragged the vowels out, or clipped them off, as compared with the templates, system performance fell. If he mumbled, performance fell. The system didn't do very well with words

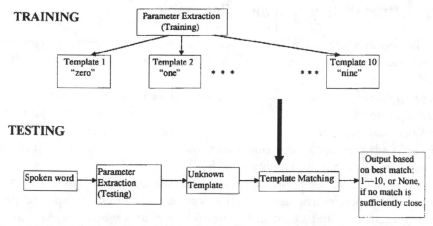

Figure 3.1 Speech recognition of digits using template matching.

outside its vocabulary. With a vocabulary of 10 words the system should have 11 responses: zero through nine and "not in the vocabulary," but the early systems responded poorly to nonvocabulary words, recognizing them in a seemingly random way.

Ideally, both training and testing had to take place in a noise-free environment. Background conversation, doors closing, or dogs barking severely limited performance.

Ideally, vocabularies were small enough so as not to contain easily confused words. Even the ten digits fail to meet this ideal perfectly—*five* and *nine* sounded very much alike to these early recognizers, as did *one* and *nine*.

Nonetheless, this was heady progress for the 1950s, and the scientists who designed and built these systems were pioneers. The limitations they encountered and documented became the subjects of research for the next forty years. (The papers in [5] are a good source of this history.)

Most of the improvements in the 1960s were incremental. Better hardware and software, and an increased interest in speech recognition on the part of academicians, led to systems with larger vocabularies and improved parameter extraction and template-matching algorithms. The major breakthrough of the decade was the discovery of **dynamic time warping** (**DTW**). This was not a way of visiting the dinosaurs. It means time normalization: The length of a test utterance was adjusted to match the length of the template with which it was being compared. DTW is discussed in detail below.

3.4.1 The Era of ARPA

In 1971, the Advanced Projects Research Agency (ARPA) challenged American companies and universities to develop a speech-understanding system with a vocabulary of at least 1,000 words capable of processing connected speech with an error rate of under ten percent in a low-noise environment for use by many cooperative speakers. The systems were allowed to have an artificial syntax and a highly constrained context and were not required to operate in real time [6].

Before we get to the actual systems that were developed, let's examine the challenge. ARPA deliberately used the word *understanding*, as opposed to *recognition*. *Understanding*, when used in this way, came to mean that once input was recognized, or partially recognized, it would be further processed. If a question was posed, the system would be required to answer it; if a request were made, the system would have to fulfill it. This working definition of *understanding* avoids the philosophical issues discussed in Section 3.2.

This kind of understanding does not require perfect recognition. Not all the words in the sentence, *"What is the height of the flagpole?"* need be recognized for the question to be answered. *What ?? height ?? flagpole* would probably suffice, where the question marks represent unrecognized sequences. An artificial syntax—rules specifying the position of parts-of-speech in an utterance—helps fill in the gaps by indicating that a question word such as *what* is always followed by a form of the verb *to be* followed by a noun-phrase expression; that, further, a noun-phrase expression is expected to be an article (such as *the*) followed by a noun (such as *height*) and possibly followed by a prepositional phrase; and that further still, a prepositional phrase is expected to be a preposition (such as *of*) followed by a noun phrase (such as *the flagpole*). All this specific knowledge about how parts of speech fit together helps the system fill in gaps left by imperfect recognition. The syntax is "artificial" rather than the "natural" syntax of ordinary English because it is far more restrictive. In everyday English, nouns, verbs, and adjectives may follow question words: *What ship is that? What leaps tall buildings in a single bound? What beautiful woman did JFK marry?* In the artificial syntax, these choices are eliminated, making it possible for the system to "understand" incompletely recognized utterances.

Similarly, a highly constrained context—one in which only certain topics may be discussed or queried—can compensate for poor recognition. If the computer mistakenly recognizes *"What is the weight of the flagpole?"* for *"What is the height of the flagpole?"* in a context in which flagpoles have a height attribute, but not a weight attribute, the question may be interpreted correctly through artificial intelligence. The computer would "reason" as follows:

> Flagpole heights are a possible topic; flagpole weights are not a possible topic; the words *height* and *weight* sound alike and may be confused by the speech recognizer. Therefore, when the recognizer "hears" *weight,* but *height* makes sense, it will be assumed that the speaker actually said *height.*

Nonetheless, the utterances had to be within range of the predetermined topics. No ARPA project speech understanding system could respond to off-the-cuff, context-free sentences such as : *"How much is one plus one?"* or *"Tell me what day of the month tomorrow is."*

The term *connected speech* also had a special meaning in the ARPA challenge. On the positive side, it meant you didn't have to speak words one at a time with a short pause between them. ARPA wanted to push beyond the isolated-word recognizers of the 1960s. On the negative side, it meant that articulatory effects were not necessarily taken into account. Thus, one had to say 'didyou' [didyū] and couldn't say 'didja' [dijə]. The ARPA projects did not require continuous, free-flowing speech.

For its time, a 1,000-word vocabulary was large. Much communication can be achieved with a judiciously chosen 1,000 words or even much less. The problem for a user is restricting one's self to that vocabulary. A vocabulary of four words will allow you to ask the time: *What time is it?* However, people have many ways of posing the same question: *Can you tell me the time? What's the time? You got the time? What's your watch say?* You need to remember the right way. Combined with a highly restricted syntax and context, however, the task of vocabulary confinement is manageable. Altogether, the restricted syntax, context, and vocabulary make up the **sublanguage** of the application.

What does the requirement for "an error rate of under ten percent" mean? It could mean that for every 100 words spoken, the system must recognize at least 90 correctly. However, in a highly constrained speech-understanding system, high recognition rates aren't necessarily needed. The system can often determine what to do based on recognizing a few key words. The ARPA error criterion meant that 90 percent of the questions or requests posed to the computer should be responded to satisfactorily. We will examine error rates and speech recognizer performance in more detail in Section 3.10.

Allowing a low-noise, and hence somewhat artificial, environment was another concession to the difficulty of the task. Most real-world environments are full of noise, and noise interferes with accurate recognition, both by humans and computers. This was yet another way in which the ARPA project ignored the exigencies of practical speech understanding in order to focus on the scientific and computational issues.

Early systems were highly speaker-dependent. They would respond best to the speaker who created the templates, which were called **reference models.** This term came to be used for the more sophisticated internal representations of word spectral patterns developed during the ARPA projects. The ARPA projects tried to move in a direction of more speaker independence when they asked that the systems be usable by "many cooperative speakers." "Many," it turned out, meant several, and these individuals still had to provide some training sentences so that the system could adjust the reference models according to who was speaking. The matter of speaker dependence/independence is one of degree rather than of dichotomy. Many factors plays a role. We will examine them more fully in Section 3.5.4.

The requirement that the speakers be cooperative is, on the surface, mundane. Who wants to deal with uncooperative people in any circumstances? Like nearly every criterion laid out by ARPA, this one had hidden and profound implications. "Cooperative" meant not only "willing to cooperate" but "willing to learn how to speak to the system," in effect, learning a new dialect. Researchers and their assistants, who often are university students, are willing and able to cooperate to this extent. The general public is not. The most you can hope for is to minimize mumbling. Worse, when speech recognition systems are seen as a threat to job security, invasive, or impersonal, speakers may become uncooperative in an effort to break the system. That would be like me becomng an uncoerative authr, say by delibrtly omitin lettrs frm wrds. Uncooperative speakers can ruin the performance of any speech recognizer ever built.

Most speech recognition applications require real-time performance. That means when the bad guys are closing in on Luke Skywalker at mach 1,000, and he instructs his spaceship's computer to "raise the deflective shields and take evasive action," the computer had better not take five minutes to process the command. Well, how about one minute? No? Would you settle for 10 seconds? Probably not, but that was ARPA's requirement: a few times real time on the fastest hardware available. It is unclear exactly what "real time" means in this context. Is it the time it takes a human being to respond after hearing the entire command, which is half a second, or is it the time it takes a human being to respond including the time spent hearing the command? I take it to be the latter based on the fact that the human being listens and processes simultaneously, and presumably the computer does too. Luke's command takes two to three seconds to speak, and that, plus a small delay for human response, is real time.

It is not unreasonable to design large real-time software systems to run in ten or twenty times real time in the development stage. There are two reasons for this. When the software is proven, its slowest components can be run on dedicated chips, with a huge increase in speed. Second, conventional hardware speed has historically increased exponentially, rising by a factor of 10 every five years. Thus, ARPA's real time requirement was not exigent.

Finally, ARPA did not require systems to be cost-effective. This is generally true of research. It's unclear as to what a cost-effective system might be, involving as it does production, marketing, and sales. Nonetheless, cost effectiveness is of concern to practical systems.

At the end of the project in late 1976, three contractors, Carnegie Mellon University (CMU), Bolt Beranek and Newman (BBN), and System Development Corporation (SDC)-Stanford Research Institute (SRI), had produced six systems. The three most viable were the Harpy and Hearsay II systems of CMU and the Hwim ("Hear what I mean") system of BBN. Of these only Harpy fully met the five year goals of ARPA. Harpy accepted connected speech from five cooperative speakers in a low-noise environment. The speakers had to utter 20 training sentences so that the system could tune itself to their speech. The vocabulary was 1,000 words and the error rate five percent. The syntax was highly constrained, as was context. Details of these and other ARPA project systems may be found in [6].

Harpy ran in about 80 times real time, but on a machine 250 times slower than the best technology available in 1976. Running times do not always scale up in exact proportion to processor speeds, but it would be safe to say that Harpy could have run in real time if the best hardware of the era had been available. It reportedly cost $5 to process a sentence with Harpy, though the cost-accounting challenges of verifying this figure are daunting. It is amusing to contemplate when the $5 cost would be worth it. To raise the deflective shields and take evasive actions (is that one or two sentences from a cost basis?) would be worth the five or 10 bucks. To dictate a business letter might not be.

As with most research, questions were among the major yields. Asking the right questions is as vital as providing the right answers. The ARPA projects' research brought the true difficulties of effective speech recognition into the light and laid bare the many challenges that had to be met.

3.4.2 After ARPA

The past two decades of speech recognition research have focused on four areas: statistical modeling techniques, utilizing higher levels of linguistic knowledge, noise immunity, and cost.

3.4.2.1 Statistical Modeling

Harpy employed a probabilistic spectral pattern matching technique based on Markov models. Variations of the method were studied and implemented in the 1980s, and today many successful speech recognition systems are based on Markov models.

On a higher level, statistical modeling of word-pair likeliness also contributed to recognition, and today the technique is *de rigueur* for large vocabulary systems. If a person said *big fat pig* and the recognizer's best guesses were *pig fat pig, big fat big, big fat pig,* and *pig fat big,* a statistical model developed from millions of English sentences would tell it that *big fat* and *fat pig* were more commonly found pairs of words than *fat big* or *pig fat.* Combining these statistical facts makes the computer prefer *big fat pig* from among the four choices.

3.4.2.2 Linguistic Knowledge

Speech recognition is not just a matter of matching spectral patterns to phonemes. Human pronunciation is inconsistent in its finer detail, and spectral patterns are by nature fuzzy. On different occasions, pronunciations of /t/ and /d/, or /i/ and /e/ overlap, so it is impossible in principle to distinguish between them based on phonetics alone.

Fortunately for us language is redundant on many levels. Only certain sound combinations form syllables. For example, no syllable in English begins with /tl/. Similar constraints are found among words, which do not combine higgledy-piggledy to form sentences. Rather, they conform to rules of syntax. If I'm missing a letter in *John _as seen the truth,* I know it cannot be *g or w* even though *gas* and *was* are common words. Only *h* produces a word that conforms to English syntax. If the sentence is *John _ed his dog last night,* semantics suggests /f/ very strongly, with the possibility of /l/ and the unlikelihood of /w/.

The ARPA project pioneered the use of linguistic knowledge. Hearsay II borrowed the "blackboard" notion from the artificial intelligence field. *Blackboard* is jargon for a database of information made available to the diverse processes of a software system. Hearsay II had various subparts that checked on whether a potential sound sequence was consistent with syllable structure, whether a potential syllable combination was a legitimate word, whether a potential word combination was a legitimate phrase, and so on. Through the blackboard, information from these various levels of knowledge sources could be exchanged. Thus, if a potential word wasn't found in Hearsay II's dictionary of allowable words, the system could back up and substitute a different, sound or syllable, forming a different word, which it could then try out.

Hwim employed a syntactic analyzer called an "augmented transition network" that eliminated phonetic choices that led to ungrammatical sentences. Harpy achieved a similar end by means of a "finite state grammar." (Both of these syntactic analyzers are described in [7]). In both systems, if the recognizer's best guess was ill-formed, say, *John green its dog,* the syntactic component would ask the recognizer for its next best guess and continue to do so until a grammatically acceptable sequence occurred. If no well-formed sentence could be found, the system rejected the input as unrecognizable. All large speech recognition systems developed after ARPA had ways to restrict recognition choices based on the syntactic constraints of the language.

3.4.2.3 Noise Immunity

Noise is ubiquitous. However, the ARPA projects were concerned chiefly with the kinds of fundamental problems we've been discussing. None worried about noise. Experiments took place in quiet environments using high-quality electronics. The quest for practical, usable systems led to an investigation of the effects of noise, which can be devastating. Systems with five percent error rates in quiet environments found themselves with 35 percent error rates when a Mozart serenade was played in the background, and if it was the *1812 Overture,* forget it.

Background noise, like the cannons of 1812, is not the only kind that affects speech recognition. Channel noise plays havoc with the recognition process as does noise introduced by the speaker such as coughing, throat clearing, snuffling, snorting, sputtering, spluttering, stuttering, stammering, slurring, lisping, lip smacking, and nonlinguistic vocalizations such as hemming, hawing, uh-ing, and er-ing.

These difficulties were addressed throughout the 1980s. Advances in electronics led to improved noise-canceling microphones. An understanding of the distortions introduced by the telephone network allowed them to be modeled and accounted for during the recognition process. Some extraneous sounds introduced by speakers could be detected and ignored during recognition. Human factors experts addressed the problem of getting users to speak fluently. In all, immunity to noise improved greatly and led to widespread applications—for example, voice dialing of mobile phones in moving automobiles—that might otherwise have not been feasible.

3.4.2.4 Cost

If you had to pay cash for an ARPA project speech recognition system you'd be out millions. The post-ARPA history of speech recognition saw the price tumble, much as it did for desktop computers. I remember my (research project's) first speech recognizer. It was a Japanese model, capable of handling a 200-word vocabulary of connected speech. It cost about $30,000 (thank you, National Science Foundation) and was considered state of the art. The year was 1982. Nowadays, the tables are turned. A model capable of handling a 30,000-word vocabulary costs $200. More modest speech recognition systems are available for personal computers for under one hundred dollars.

3.5 Three Dimensions of Speech Recognition

Speech recognition systems are classified and priced a little like automobiles. Your basic car has a stick shift and an AM radio and no air conditioning or power windows. Your basic speech recognizer only accepts speech spoken with pauses between each word, must be pretrained to your voice, and is limited in vocabulary. A few more dollars will get you an automatic transmission or stereo system in your new car. Likewise, coughing up some dough will get you a speech recognizer that lets you run words together and may recognize the speech of your friends if their voices are similar to yours.

3.5.1 Continuous Versus Noncontinuous

The first speech recognizers only accepted discrete speech, which meant that the words had to have pauses of several hundred milliseconds between them. Actually, the discrete units didn't have to be words, and often weren't. They could be utterances of several words spoken without pauses. The recognizer would treat such utterances as if they were words. I'll continue to talk about "discrete word" recognizers when "discrete utterance" recognizers is what is meant.

The reason for the pauses was to make word boundaries easy to identify. It's essentially the same reason we leave a bit of whitespacebetweenwordsona—whoops—printed page. It's a whole lot easier to figure out what the words are. Pauses had to be long enough so as not to be confused with the 60 or so millisecond moment of silence that precedes the release of a stop consonant.

The disadvantages of discrete input are unnaturalness and slowness. It is natural, even when reading a discrete list of numbers, to run the words together. As for speed, a normal pace for most people is around 150 words per minute. With one-third of a second pauses between words, it would take one minute and 50 seconds to speak 150 words, an effective entry rate of 82 words per minute. In practice, even skilled speakers have trouble maintaining 60 words per minutes because the method is so demanding. Although it is claimed that speakers can achieve a higher rate, my own experimentation has indicated otherwise [8].

Speech recognizers capable of accepting **connected speech** appeared on the market in the early 1980s. *Connected* means that the words are fully pronounced but that there need not be pauses between them. To see how tough it is to segregate connected speech into words when you don't know the language, even with occasional comma pauses, try this (assuming you don't know Portuguese):

Apossedalinguagem, maisdoquequalqueroutroatributo, distingueosseres humanosdosanimais.

Well, what's taking you so long? Connected speech recognizers had to perform extensive computations to sort out the combinatorial possibilities of what sounds, syllables, and words were represented by long, unbroken spectral patterns. The early recognizers were trained extensively on pairs of words, focusing on their boundaries, so later they would recognize where words came together. However, even as few as 100 words create 4,950 word boundaries when paired up, a number that grows exponentially with vocabulary size, so it becomes impossible to prestore and later recognize every word boundary.

Connected speech is tedious and unnatural because it forbids the natural process of running words together. **Continuous speech** covers the ground from connected speech with few interword articulatory effects—didja-eat-yet—through massively run-together speech—jeet-jet, for *did you eat yet.* It is common in the field to play fast and loose with the term *continuous speech.* Everyone wants it, but no recognizers are capable of handling it in its untamed form.

A kind of recognition called **word spotting** is a blend of discrete and continuous. The recognizer accepts continuous speech as input, but seeks to recognize particular key words embedded in the speech.

3.5.2 Speaker-Independent Versus Speaker-Dependent

The early speech recognizers were speaker-dependent. Unfortunately, many speech recognition applications require speaker independence. They need to recognize the speech of persons who did not make reference models.

Even the primitive recognizers of the 1950s were not entirely speaker-dependent. If John was a 200-pound male baritone who trained the machine, a different 200-pound male baritone speaking a similar dialect could use it with some success, although a 120-pound female soprano probably could not.

The chief factors controlling speaker dependence are the personal voice characteristics of the speaker and the dialect spoken. Everybody's voice is somewhat different, to a greater or lesser degree. If person X trains a system, and its reference models are derived entirely from that one source, than recognition success will decline when persons other than X attempt to use the system, and the decline will be more precipitous when the personal differences are greatest. If X is male, females will have much less success than other males, all things being equal. If X is an adult, children's voices will not be recognized very well.

Humans are highly speaker-independent with regard to personal voice characteristics. The ability to adjust to different voices appears to be part of our innate linguistic capabilities. On the other hand. humans are less capable of adaptation when different dialects come into play. A westerner will not recognize the speech of a southerner as readily as another southerner, other factors being equal. In general, the farther apart the dialects, the worse the recognition.

This situation is reflected in speech recognizers. Certain large-vocabulary, discrete word speech recognizers such as those used to take general dictation are ostensibly speaker-independent. They come with reference models in place, and users do not have to create them. However, they adapt dynamically to the particular voice characteristics of the user. When a person first begins to use such a system, performance is poor, but as usage time grows, performance improves. Such systems actually become more speaker-dependent over time.

With dialects, if a Bostonian trains a speech recognizer to the word *data* by pronouncing [dātər], and a midwesterner tries to have it recognize *data* by uttering [dātə], failure is likely. Americans traveling in the United Kingdom and Ireland often meet people whom they can barely understand, though they speak English. The speech recognizer is in the same boat when the person using it speaks a dialect different than the person who trains it. Adaptation to different dialects is harder than adaptation to personal characteristics, for recognizers as for humans.

The term *speaker independence* is bandied about by speech recognition marketers, but it is not well defined. A speech recognizer certainly isn't independent of speakers who don't speak its language, who have speech defects, who speak an unusual dialect, who have idiosyncratic speaking habits, or who are uncooperative. It's a matter of degree.

3.5.3 Vocabulary Size

The third dimension of speech recognition is vocabulary size. It's what you think it is: the number of words (or utterances) to which the recognizer is trained, and for which it has reference models for future matching.

In the beginning, vocabulary sizes were small, 10–20 words. A typical vocabulary before ARPA was the eleven digit-words *zero, oh, one, two, three, four, five, six, seven, eight, nine, ten,* plus a few control words such as *enter, stop, go, backup, repeat.* After the ARPA projects vocabularies in the thousands of words became feasible. Speech recognizers developed in the 1980s and 1990s had successively larger vocabularies, some as large as 100,000 words.

With vocabularies of any size, confusion is possible. The larger the vocabulary, the more opportunities for confusion, but size is not the only factor. A 26-word vocabulary consisting of the letters of the alphabet will be more highly confusable than a similarly sized vocabulary of names of states of the United States. That's because the letters contain confusable sets, for example *bee, cee, dee, ee, gee, pee, tee, vee, zee.* Also, letter names are one syllable long. The shorter a word, the more difficult it is to recognize because less spectral information is available. The names of most states of the United States are polysyllabic and do not rhyme with one another.

Vocabularies may contain various pronunciations for the same word. The word *the* is alternatively pronounced [dhə] and [dhē] by English speakers, so it counts as two vocabulary words for the speech recognizer, though you and I would say it's one word.

Multiple forms of the "same" word require different reference models in a speech recognizer. Thus *recognize, recognizes, recognized, recognizing, recognizer, recognizers, recognition, recognizable, recognizably, recognizance, recognizant, recognitive, recognitory*—all with slightly different pronunciations—would be entered separately and counted as 13 words when preparing marketing information, though dictionary makers would cover the 13 words in three entries: *recognition, recognizance,* and *recognize.* Thus, the terms *vocabulary word* and *dictionary entry* mean different things.

In a similar vein, homophones—different words pronounced the same such as *I, eye,* and *aye*—need have only one reference model, though they may count as several vocabulary words.

The act of dictating a business letter to a large vocabulary speech recognition system requires more than merely reading the letter (in discrete speech). Whenever the system is unable to decide definitively on a word, it stops and forces you to choose from among its best guesses. You might be saying: *We recently shipped twelve cranes . . .* when the system falters on *cranes* and up pops a menu like:

1. Cranes;
2. Brains;
3. Drains;
4. Grains;
5. Trains.

You must then stop and choose the intended word, either by pointing with a mouse, or speaking the number beside the word. The larger the vocabulary the more frequent such interruptions become, so requesting and paying for a larger vocabulary than needed may have a negative effect on both efficiency and cost.

3.5.4 Tradeoffs and Interactions

Each of the three dimensions can be measured along a sliding scale, in spite of the dichotomous usage of terms like *discrete speech* versus *continuous speech,* or *speaker-dependent* versus *speaker-independent.*

Discretely spoken speech with one-third of a second pauses between words represents one end of the type-of-speech scale. The other end is rapidly spoken, highly run together speech. Between the two extremes is continuous speech with lesser degrees of interword articulatory effects and discrete speech with smaller and smaller pauses between words. When the pause duration reaches zero, we call that connected speech.

The speaker dependency scale has at one extreme a system trained and used by a single speaker. The other extreme is usage by every mother's son and daughter that speaks the language or languages for which the recognizer is trained. There are today multilanguage speech recognizers. Many of them are sponsored by NATO and feature the languages of NATO members (Danish, Dutch, English, French, German, Greek, Icelandic, Italian, Norwegian,

Portuguese, Spanish, and Turkish.) There are also recognizers for Japanese, Mandarin Chinese, Hindi, and other languages of the world. Thus, speaker dependency is two scales in one, because individuals may differ along a sliding scale of personal voice characteristics, and another scale of dialect and language differences. We assume that there is some way of combining the two scales into one, perhaps by some kind of averaging.

There are many ways to achieve partial speaker independence. One is to have each potential user train his or her own set of templates or reference models. These can then be averaged so that any one of the trainers or persons similar to them could, with limited success, use the system. This method can be carried to various extremes, and has been. Thus, for training small vocabularies, exemplars for each word can be collected from hundreds of voice types and dialect regions, averaged, and then made to work on entire populations.

It is also possible for each of several users to have their own personal reference models, which the speech recognizer loads when that person takes the microphone. This is a way of simulating speaker independence with a machine that is speaker-dependent.

Vocabulary, of course, is on a sliding scale of number of utterances for which reference models are created. At one extreme would be a vocabulary of two words and the other extreme is open-ended.

Many factors ultimately affect the performance of a speech recognizer. The cooperation and skill of the speaker, control of both background and channel noise, the algorithms that are implemented, the hardware in use, and so on. Measuring performance is not a trivial task. (We'll consider it in Section 3.11.) However, whatever it is, and however we measure it, the following is true: As you move along the sliding scales toward continuous speech, or toward speaker independence, or larger vocabulary sizes, performance will degrade. Conversely, as you slide toward discrete speech, speaker dependence, or smaller vocabularies, performance will improve.

The three dimensions interact. You can compensate for a large vocabulary size by demanding only discrete speech. Until recently the large vocabulary recognition system of tens of thousands of words only accepted discretely spoken speech. Current systems accept connected speech but woe be unto the user who drifts into the dismal swamp of continuous speech with articulatory deformation. The large vocabulary system trains itself to a particular user's voice—a process that occurs during usage and may take several weeks to produce improvement. In effect, the system becomes more speaker-dependent, which is needed to compensate for the large vocabulary size.

The difficulties of continuously spoken speech with modest interword effects may be partially obviated by reducing the vocabulary size. There are systems that perform well on inputs of rapidly spoken streams of digits. They perform even better when tuned to an individual speaker—that is, moved in the direction of speaker dependence.

A high degree of speaker independence is best achieved on, you guessed it, small vocabularies and discretely spoken speech. In the field, speaker independent systems generally have vocabularies of digits alone, to be spoken if not discretely, then with minimal slurring between words.

If we create a three-dimensional space and assign one of our sliding scales to each dimension, we can visualize the interaction by observing that performance is more or less inversely proportional to the distance from the origin in the three-dimensional space. This is suggested in Figure 3.2.

3.6 Units of Speech Recognition

All computer speech recognition consists of training, in which the system is exposed to known speech from which reference models are constructed, and

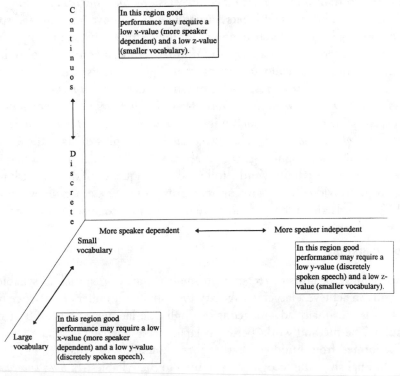

Figure 3.2 Trade-offs in speech recognition systems.

usage, in which models of unknown speech are compared to the reference models in an attempt to find a close enough match for recognition.

Reference models may be based on units as large as a phrase, or as small as a fraction of an allophone, with various choices in between. In this section, we examine the ramifications of the choice of speech unit.

3.6.1 Words and Phrases

Early speech recognizers used words, or several words spoken together, as units. For example, the utterance *go back* might be entered as a single unit, rather than having two words *go* and *back,* neither of which was used elsewhere. Also, the longer an utterance the easier it is to distinguish.

Applications designers quickly learned that the little words were difficult to recognize, in part because people give their pronunciations short shrift, especially in continuous speech. For example, if an application was to use the expressions *the cook, the wife, the thief, the lover, a cook, a wife, a thief, a lover,* logic might dictate training the six words *cook, wife, thief, lover, the, a.* But count on it, in practice the *the's* and the *a's* will often not be recognized correctly. Rather, it is better to train on the eight longer expressions *the wife, the thief, . . .*

Clever applications designers learned how to improve "continuous" speech recognition by having reference models for long expressions. For example imagine a speech interface to a database of information about your rock collection in which specimens are catalogued using four-digit numbers. The speech recognizer might be expected to respond to questions such as *what is the weight of rock 2-3-4-5? what is the composition of rock 2-3-4-5? where are the samples containing lapis lazuli?* The reference models might be *what is the weight of rock, what is the composition of rock, where are the samples containing,* plus digits and the names of minerals. Thus recognizing *what is the weight of rock 2-3-4-5* is a matter of recognizing one very long "word," followed by a sequence of digits, a task much easier than individually recognizing six words plus digits, especially when three of the six words are little words that are difficult to recognize (*is, the,* and *of*).

3.6.2 Syllables

The advantage of units of recognition smaller than words, such as syllables, is that you can achieve a larger vocabulary for a smaller number of reference models. This is minimally advantageous in English, a language of 10,000 or more syllables. The method works better in Japanese, a language in which all words can be formed from hundreds, rather than thousands, of syllables.

In English, if the recognizer is trained to the syllables *-ly, bad, poor, worth, -less, god,* and *man,* the words *bad, badly, poor, poorly, worth, worthless, god, godly,* *godless, man, manly,* and *manless* can be recognized—12 words for the price of seven reference models. The problem is that suffixes such as *-less* tend to be pronounced with a schwa in words like *godless* [godləs] but pronounced with a full-valued vowel in isolation [les], and this is true of other syllables. Thus, when training *-less* a person tends to say [les], but when using it in a word it emerges as [ləs]—tricky, but not impossible for the recognizer to figure out. It's all due to the nasty little habit that English has of using a schwa [ə] as a general, all around unstressed vowel.

The second problem has to do with intersyllable articulatory effects. Suppose the syllables *gen,* and *tle* [təl] are in the training set. From these two syllables you would hope to recognize *gentle.* Unfortunately, most (American English) pronunciations of *gentle* omit the *t,* so phonetically the recognizer hears [genəl].

Although syllable-based recognition is more appropriate for Japanese, the same problems of disappearing vowels and articulatory degradation also occur, as they do in all languages. The Japanese word *sukiyaki* is formed from the syllables *su, ki, ya, ki.* When the word *sukiyaki* is spoken in a freely flowing sentence, however, the *u* is elided entirely, the *s* is lengthened to compensate for the missing *u,* and the first *i* is collapsed into the [y] so that the actual pronunciation in continuous speech is [s:kyakē]

3.6.3 Phonemes

An even larger words-to-reference models ratio can be achieved by training smaller units on the level of individual sounds. Although this is commonly called "phoneme-based recognition" the term is a misnomer, one I hope readers of this book (especially Chapter 1) will appreciate. It should be called "**allophone**-based recognition" since allophones are what we call sounds as actually pronounced.

It is impossible to count the allophones of a language. In some sense, there are infinitely many. However, we can identify cardinal points in allophonic space such as [pʰ] and [p], allophones of the phoneme /p/. Although there are infinitely many degrees of aspiration, we simply distinguish between its presence and absence, an oversimplification but one that works in practice. Similar remarks apply to all phonemes and their allophones, both consonants and vowels.

A set of around 100 allophones will suffice for all the words of English. We make reference models for 100 allophones and proceed to recognize speech in essentially the same way we read.

The actual production of reference models is less straightforward than asking users to train the system by speaking words or syllables. How does one speak the allophone [p]? Instead, samples of speech that contain the sounds of interest are collected, isolated, and then modeled. This is easier said than done. For one thing it's difficult to know which pronunciations to choose as the ideal representation of a given allophone. Second, it is not clear from the spectral patterns where an allophone begins and ends.

Many of the same difficulties plague this choice of speech units as other choices. Cheshire cat-like vowels pop in and out of existence, and when they're there, their value varies. For example, *chocolate* may be pronounced phonetically in dozens of ways. Here are eighteen of them:

chôklət	chôklit	chôklet	chôkələt	chôkəlit	chôkəlet
chôkōlət	chôkōlit	chôkōlet	choklət	choklit	choklet
chokələt	chokəlit	chokəlet	chokōlət	chokōlit	chokōlet

In continuous speech this problem would be compounded and could—and does—lead to a combinatorial explosion.

3.6.4 Diphones and Triphones

In an effort to avoid interallophonic articulatory effects, **diphones** were used as reference models. A diphone is the second-half of one allophone concatenated with the first-half of another allophone. Thus the center of a diphone is the boundary where two allophones abut, permitting interallophonic phenomena to be modeled. The number of reference models climbs precipitously. One hundred allophones would produce 4,950 diphones if all combinations of pairs are considered. Still, it's not a bad deal: all the words of English for the price of some 5,000 reference models.

Triphones take a somewhat different tack. They are based on the notion that allophones, like a good story, have a beginning, a middle, and an end. The middle is relatively invariant. The two endpoints vary with context and are the locus of interunit articulation effects. The number of different triphones is staggering. They have to be collected automatically from large speech databases. If the databases were created from the speech of a wide variety of speakers and dialects, triphones suitable for a high degree of speaker-independent recognition can be accumulated.

When the vocabulary for a given application has been determined, vocabulary reference models can be constructed from the triphones in hand, rather than from individual speakers. A speaker adaptation algorithm adjusts the reference models of the triphonic vocabulary words to the idiosyncrasies of particular users. For a large vocabulary, it may take weeks or months to complete the adaptation process. These systems are intended to be used over a long period of time by the same speakers.

Although the use of diphones and triphones as units of speech recognition have made large-vocabulary, speaker-independent systems more viable, the same problems of speaker variation, dialect variation, and the articulatory degradation that normally accompanies continuously spoken speech still prevent performance from being good enough for many potential applications.

3.7 Representing the Units

Representation is arguably the most challenging problem in computer science. Given zeros and ones, how to do you represent numbers, letters, bank transactions, chess games, faces, speech, . . . inside the computer?

In its rawest form, speech is represented as a sequence of numbers, the result of the A-to-D process, itself hardly straightforward, as you know from Chapter 2. This form requires a lot of refining. One word alone, perhaps 350 ms of speech, may consist of some 3,000 samples and consume 24,000 bits using an eight-bit quantizer. Even a tenth that number of bits—2,400 per word—is untenable in raw form for use in speech recognition. With no further processing, the bit patterns of two pronunciations of the same word on different occasions are no more likely to show similarity than two pronunciations of different words.

The refining process may be as simple as a bit reduction algorithm that attempts to extract something unique about the pattern of bits that represent a speech unit. Or it may be a complex process in which parameters of speech are deduced from the acoustic features of the speech signal and used in model construction. In Section 3.7.1, we'll look at some ways of defining and measuring acoustic features to model speech.

3.7.1 Acoustic Features

Speech patterns do not reveal themselves very well in the time domain. The frequency domain, where spectral characteristics are more manifest, is where speech recognition takes place. The earliest speech recognizers used banks of filters to divide the spectrum into regions based on frequency and then extracted parameters from each region. You can see where this might work for vowels. Parameters proportional to intensity would be greater in frequency bands corresponding to the formants and thus be related to vowel quality.

In practice the spectrum is examined every five milliseconds in chunks or **windows** that are 20 ms wide. (Of course, these numbers may vary somewhat, depending on the particular method of analysis.) The overlapping windows (see Figure 3.3) provide a measure of redundancy that improves the accuracy of spectral features at the cost of additional processing time. For each window a vector of parameters is measured and stored. A collection of these vectors over the duration of the speech unit represents the essential characteristics of that unit.

3.7.1.1 Vector Quantization

The coefficients of LPC may be used as parameters for characterizing speech. As we saw in Chapter 2, these coefficients are particularly well representative of formant location, hence vowel quality. They have the further advantage of computational efficiency. Cepstral coefficients, which are parameters drawn from the cepstrum of the quefrency domain, have also proved to be effective means of representation for speech recognition. Other parameters are also discussed in the literature on speech recognition, including a number of minor variations of LPC and cepstral coefficients. (Many of these details are discussed in [9].)

All of these parameters have the advantage of reducing the speech unit to its essential defining characteristics. Although the bit reduction is substantial when compared to the raw data in the time domain, it is still not enough for many purposes. To reduce further the number of bits needed to represent a speech unit, a technique called **vector quantization (VQ)** may be employed.

Suppose each 20-ms window is represented by a vector of 10 numbers. They may be LPC or cepstral coefficients, or something else. Each vector can be viewed as a point in 10-dimensional space (whoa! don't try to picture it), just as a vector of two parameters can be viewed as a point in a two-dimensional plane. As it turns out, these points in 10-dimensional space are not equally dispersed. Rather, they occur in clusters. Vector quantization identifies these clusters and assigns a single number to each one. The process is illustrated in Figures 3.4 and 3.5. In Figure 3.4 I've shown 16 points in two-dimensional space. You may imagine that each point represents some acoustic properties of some speech unit.

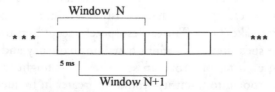

Figure 3.3 Overlapping 20-ms windows taken at 5-ms intervals.

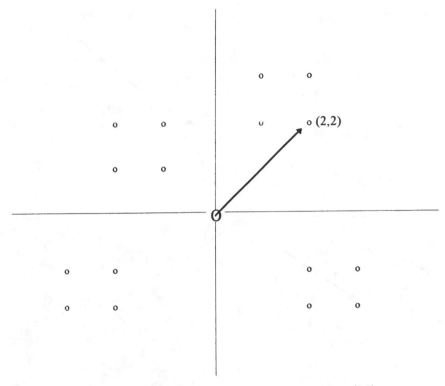

Figure 3.4 Sixteen points in two-dimensional space, showing the vector (2,2).

Each point may be interpreted as a vector, which is indicated in the figure connecting the origin to the point (2,2). The points are clustered into four groups. In Figure 3.5, I've averaged each cluster and indicated the average point, called the **centroid**, with a little square. It's determined in such a way as to minimize the difference, or error, between it and all the other points that go into its determination. Each little square, of course, may be interpreted as a vector in the plane, as is shown. Finally each vector is arbitrarily assigned a number, starting with zero.

The four vectors are described by their (x, y) coordinates, so we could make a table in which we put the vector's number in both decimal and binary notation along with its values. (If we were in 10 dimensions, each vector would have 10 values.) See Table 3.1.

What we have achieved here is the ability to approximate a vector using two bits, as opposed to representing it exactly using two numbers, each of which may require several bits. We have some overhead cost in the form of the table, but it is comparatively small because in practice it is reused multiple times. The universal principle of "no free lunch" applies in that vector quantization introduces

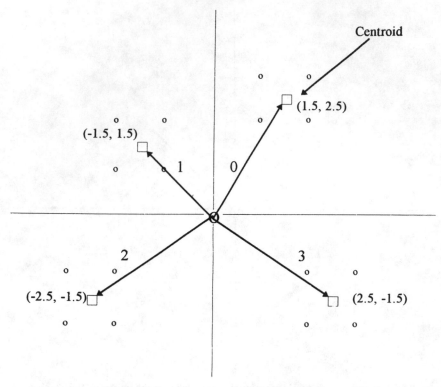

Figure 3.5 Sixteen points in two-dimensional space separated into four clusters with centroids.

error, as do all approximations, and much attention has to be paid to the methodology for reducing that error in practice.

In the jargon of speech processing, Table 3.1 is called a **VQ codebook;** the vector number is called the **codebook entry;** the vector value is called the **codebook value.** The relabeling is shown in Table 3.2.

Table 3.1
Vector Numbers and Their Values

| | Vector Value | |
Vector Number (Bits)	X	Y
0 (00)	1.5	2.5
1 (01)	−1.5	1.5
2 (10)	−2.5	−1.5
3 (11)	2.5	−1.5

Table 3.2
VQ Codebook

Codebook Entry	Codebook Value	
	X	Y
0 (00)	1.5	2.5
1 (01)	−1.5	1.5
2 (10)	−2.5	−1.5
3 (11)	2.5	−0.5

VQ is a compression technique. It may be used to gain efficiency in the transmission of speech data, it may be used in some stage of setting up reference models in a speech recognition system, it may be used to store speech data more efficiently, and so on. In all cases, a VQ codebook can only be produced from actual speech data using specific parametric representations, for example LPC coefficients. In jargon we say the codebook must be trained. Training consists in observing a wide variety of speech data so that a good sense of clustering can be obtained. Once training is over and the codebook established, all further speech data can be referenced by sequences of codebook entries.

In the example we've been working with we managed to train our codebook on 16 points. With the codebook established, if we see a sequence of acoustic data in the form of the vectors [(2.5,0), (−1,−1), (−1,−2), (1,2), (0,1) . . . we can represent it with codebook entries [3, 2, 2, 0, 1], which are the nearest codebook vectors, or, equivalently, the vectors of minimal error. The savings in bits is much greater than it seems when we look at the individual numbers that comprise the vector. Dedicating a relatively low eight bits to each number of acoustic data, we are actually achieving a compression ratio of eight to one since a pair of numbers—16-bits worth—is represented by two bits.

Here is another way to see the economy of VQ. A typical speech recognition system will digitize speech at a rate of 10,000 samples per second and use a 16-bit quantizer. Each second of speech would require 160,000 bits in the time domain. In the frequency domain suppose we use ten LPC coefficients per 10-ms window and dedicate 16 bits to each coefficient. Each second of speech would still require 160,000 bits.

Now let's build a very large codebook of 1,024 entries. It's in 10 dimensions; each codebook value will be a sequence of 10 numbers, namely the LPC coefficient values. We are now capable of representing the acoustic properties of 10-ms windows with a number between 0 and 1023, which can be done using

10 bits. Now one second of data can be represented in 1,000 bits, a 16 to one compression ratio.

For the purpose of speech recognition it is possible to associate a speech unit with each codebook entry. A sequence of codebook entries for a speech token becomes a sequence of phonetic units that correspond to a spoken word.

3.8 Comparing the Units

No matter how faithfully we represent the units of speech, if we do not have a good way of measuring their degree of likeness we will fail at speech recognition. If everyone spoke identically, like so many carefully programmed androids, speaker-independent speech recognition would be a cinch. If one person spoke identically on every occasion, speaker-dependent speech recognition would be a cinch. Unfortunately, variation is the norm, both **inter**speaker and **intra**speaker.

Much of the variation that makes comparison difficult can be eliminated by a judicious choice of parametric representation. For example, one source of variation is pitch, or the basic vibration rate of the vocal cords. People, like opera singers, are bassos, tenors, altos, and sopranos. Their vowel formants follow this pattern. The first and second formants of a basso's [i] may be 300 Hz and 1,600 Hz and that of a soprano's [i] in the range of 550 Hz to 1,900 Hz. What identifies the vowel quality [i], however, is the difference between the formants, and that is relatively independent of the pitch. By choosing parameters that reflect both the differences between formants as well as their absolute values, some pitch independence may be achieved.

3.8.1 Dynamic Time Warping (DTW)

Time duration is the speech unit feature that is most likely to make comparisons misfire. Different utterances of the same word or phrase will differ in length, sometimes dramatically. This is especially true in connected or continuous speech, where many factors govern how quickly people talk. When the speaker is excited or rushed, speech rates go up. When the speaker is being deliberate, talking to a foreigner, or simply choosing words carefully, the speech rate goes down. In both cases, the rate, hence the duration of speech units, may differ from that used when training the reference models.

Let's look at a contrived example using letters to gain a sense of the problem. Suppose a reference model of the word *labrador* looks like this: *llaabraadoorr*. When using the recognizer a speaker may utter the first part of the word leisurely, and the second part rapidly: *lllaaabrraaador*. In attempting to match these we start by matching l-l, l-l, a-l. The a-l is a mismatch so we try to improve the situation by considering surrounding letters and matching them if possible. In this case, we realize that the third l of the utterance matches the second *l* of the template. We accept this match, so we have l-l, l-l, l-l. In effect, we have compressed the beginning of the utterance to force a match. We now match up two *a*'s and then get a mismatch that can also be resolved by looking at surrounding letters. At the end of the process, the final r of the utterance ends up matching the final two *r*'s or the reference model. This, in effect, is stretching the utterance to achieve a match. This process is illustrated in Figure 3.6 and captures the essential idea behind DTW.

DTW always attempts to find a match by distorting the utterance, but sometimes this is not possible, and where it isn't DTW attempts to minimize the overall mismatch or error. In the artificial illustration in Figure 3.6, the match was perfect. If the user had said *laboratory,* instead of *lllaaabrraaador* and DTW was applied we would get the result illustrated in Figure 3.7, where several mismatches are noted.

In actual speech recognition, rather than two possible scores, *match* and *mismatch,* there is a gradation of possibilities. The process is further complicated when applied to continuous speech because most word boundaries aren't detectable. In such a case, certain limitations have to be placed on just how much time warp will be allowed to maximize the match score. For example, the

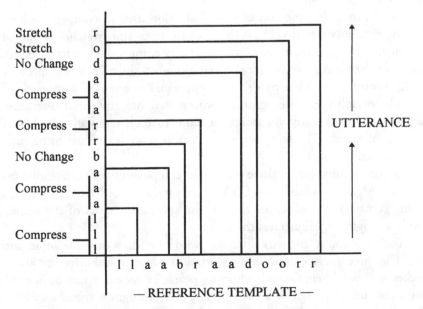

Figure 3.6 DTW using letters to represent windows of speech.

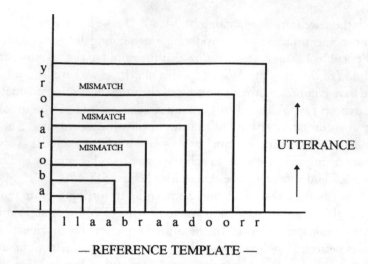

Figure 3.7 DTW using letters to represent windows of speech with several mismatches.

utterance *see yet* might score quite high when compared to the template *sit* if enough compression is applied to the *ee ye* of *see yet,* whereas if the amount of compression were limited, that error might be avoided.

In using DTW for speech recognition, in the place of letters you find acoustic features of the speech signal in spectral windows, as suggested (very abstractly) in Figure 3.8.

DTW, then, is a method of time distortion used to compensate for the differing utterance lengths. DTW both compresses and stretches in the time domain and does so nonlinearly. That is, it may compress one part of an utterance more than it compresses another part, as when three *l*'s are compressed to two (2/3 compression) but two *r*'s are compressed to one (1/2 compression).

The overall speech recognition process matches any input utterance to every reference template and assigns a score to each template based on the goodness of match. The template with the best score is deemed to be the utterance actually spoken.

In continuous speech, there may be several possibilities with similar overall scores. Algorithms based on DTW may produce **word lattices,** which are varying combinations of words that fit the spectral patterns of the reference templates. Figure 3.9 illustrates the point.

Each possibility suggests different word boundaries for the input utterance. The three guesses could be submitted to a higher authority, say a syntax checker, that would eliminate "reckon ice peach." Context would be needed to distinguish between "wreck a nice beach" and "recognize speech." We'll see more about "higher authorities" later on in Section 3.13.2.

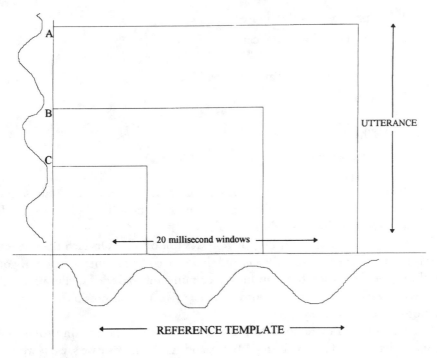

Figure 3.8 DTW using spectral characteristics. The points A, B, and C are determined by DTW in such a way as to maximize the match of the utterance window to the reference template window.

3.8.2 Hidden Markov Models (HMMs)

The word *probably* is the 208[10] most commonly used word in spoken English [10]. That may not sound like much, but it's competing with some very popular words such as I (the most popular), *and, the, to, that, it, of, a, so, on.* Not a day, probably not an hour, goes by in which you don't hear a sentence with *probably:*

W	R	E	C	K	A	N	I	C	E	B	E	A	C	H
R	E	C	K	O	N	I	C	E	P	E	A	C	H	
R	E	C	O	G	N	I	Z	E	S	P	E	E	C	H

Figure 3.9 A word lattice showing three possible combinations of words that fit the same spectral patterns.

1. "Honey, what's that noise?"
2. "Oh, it's probably the cat."

1. "Are you going to Paris this year?"
2. "Probably not."

1. "Who'll win the game?"
2. "It probably won't be us."

HMMs attempt to quantify the speech recognition notion that a certain sound in a certain context "is probably a [p], but it could be a [b], and we can't rule out the possibility that it might be a [d]."

Let's play a game. Consider three boxes full of tiles. On each tile one letter is printed from among *b,d,g,a,o,* and there are equal numbers of each. If you pull one tile from each box you have a certain probability of spelling a word, say *dog* or *ado.* Of course, nonwords, such as *ggd,* are also possible. All combinations are equally likely.

We can train these boxes to prefer a certain word and certain pronunciations of that word. Call the boxes A1, A2, and A3. Whenever we say *dog* we discard one tile of every letter but *d* from A1, one tile of every letter but *o* from A2, and one tile of every letter but *g* from A3. Other words have no effect. We repeat this process multiple times.

After the training stage of discarding tiles, if we randomly select a tile for observation (not discarding) from each box, the probability of spelling *dog* will be higher than the probability of other letter combinations. We could also say that these three boxes, when used as indicated, are a model of the word *dog.*

Suppose we pronounce *dog* in two ways, [dog] and [doog], where the doubled phonetic "o" indicates greater vowel duration. (Recall that the square brackets are phonetic alphabet spellings from Chapter 1.) Further suppose we pronounce each variant half the time when training the boxes. To model the two pronunciations best, we must make a notation on box A2 indicating that when we are selecting tiles for observation, half the time we should examine two tiles from A2 instead of one. Furthermore, suppose that about one time in ten we stutter on the *d* of dog, pronouncing it "d-dog." This tells us to adjust the model so that we observe two tiles from A1 every tenth time, on average. We illustrate the model in Figure 3.10.

In the jargon of speech recognition, the boxes are called **states;** the arrows connecting them are called **transitions;** and the tiles that are selected for observation are called **outputs.** The transition arrows indicate the probability of going to whatever state they're pointing at. When in state A1, after observing a

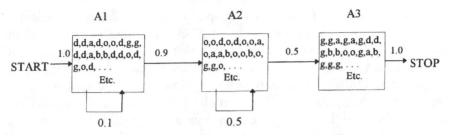

Figure 3.10 Boxes containing letters trained to prefer the word *dog* and recognize variant pronunciations.

tile, you proceed to state A2 nine times out of ten because the probability of that state transition is 0.9. One time in ten you remain in state A1 and observe another tile In state A2, after observing a tile, you have an equal probability of remaining in that state or going to state A3. After observing a tile from A3 you must stop.

Although the boxes prefer "dog," it is a probabilistic matter and other combinations are possible. This is a **two-stage probabilistic process** in that there are probabilities associated with tile observation and probabilities associated with box selection.

We can repeat this process with more boxes. For example we can train the word *bad* just as we trained *dog*, associating *bad* with the boxes B1, B2, and B3. To account for variant pronunciations of *bad, baad, baaad, baaaad* we can adjust the state transition probabilities leaving B2 to make remaining in B2 and possibly adding an *a*—more likely.

The A-boxes now model *dog*, and the B-boxes model *bad*. Let's combine the models into an elementary word recognizer, using a pattern matching, statistical technique. Suppose we have an unknown word, X. We randomly select and observe tiles from the A-boxes as described above, repeating the process a very large number of times. (This corresponds to the statistical, probabilistic nature of the process.) We record the percentage of time that X is matched by the observed tiles. We repeat the process using the B-boxes, again noting the percent of matches.

If the matching percentage from the A-boxes is sufficiently greater than that from the B-boxes and exceeds some threshold above zero, then we may conclude that X is *dog*. Conversely, a high percentage of matches from the B-boxes indicates that X is *bad*. If both percentages are below the threshold, we may conclude that X is neither *dog* nor *bad*.

This system is, in essence, a simple **hidden Markov model**. It's "Markov" because that's the name given to state transition models in which the next state is determined solely from the current state. It's "hidden" because the actual state

sequences are concealed from us. For example in observing tiles, it's possible that
d-o-o-g comes from the state sequence A1-A1-A2-A3, though it is more likely
to come from A1-A2-A2-A3. All we know is that the output is d-o-o-g and we
can only infer probabilistically which set of boxes corresponds to the input.

In using HMMs for actual speech recognition, each "tile" corresponds to
a vector of acoustic parameters. These may be derived from LPC or cepstral coefficients; they may be codebook entries from a VQ process; or they may be
something else. Each vector has a probability associated with it, analogous to the
different distributions of letters in boxes. The probabilistic nature of these parameters helps account for the variability in pronunciation in a way analogous to
how we did it with spelling ([dog] vs. [doog] vs. [d-dog] vs. [d-doog]). The state
transition probabilities accomplish the effect of time normalization by allowing
the system to remain in the same state, corresponding to a lengthier pronunciation of some speech unit. It is also possible for states to be skipped, thus ac-
counting for pronunciations in which sounds are omitted, such as the second *e*
in *federal.* A more general form of the model appears in Figure 3.11.

In use one begins at state S_0 and proceeds to either S_1 or S_2 with probabilities P_{01} or P_{02}. Whenever you arrive at a state, a vector of acoustic parameters is output (or "observed") according to certain probabilities. Thus if you
come to S_1 you get vector V_{11} with probability B_{11}, vector V_{12} with probability
B_{12}, and so on. The sequence of vectors that is generated as you proceed from
Start to *End* represents the acoustic properties of the triphone. The probabilistic nature accounts for the fact that on different occasions, the same triphone

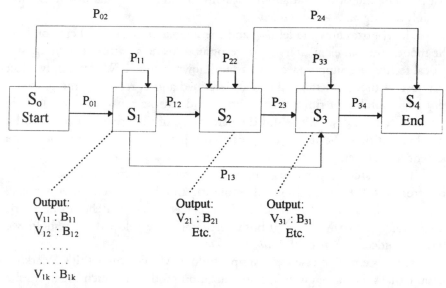

Figure 3.11 A Markov model appropriate for triphone recognition.

may have slightly different acoustic properties. For example, the onset of a triphone may be omitted in continuous speech. That eventuality is accounted for by allowing S_1 to be skipped with the probability P_{02}.

Each model is individually trained with multiple repetitions of its triphone, tuning the probabilities so that later recognition will be accurate. This is a long, computationally expensive process that fortunately need only be done one time off-line, although occasional online tune-ups may be necessary. The recognition process is essentially one of matching the sequence of acoustic parameter vectors of the input to the various outputs of the various models. The "winning" triphone corresponds to the model whose outputs best matches the input, providing the match exceeds a threshold. (Otherwise "no match" is declared.)

HMMs are the most common type of reference model in use today. Their two-pronged probabilistic nature makes them extremely effective for representing speech, since allophonic distribution is statistical in nature. From a mathematical point of view, HMMs are well-behaved and thoroughly understood. Computationally, many algorithms have been devised for carrying out both training and recognition as efficiently as the mathematics allows. An excellent mathematical description of HMMs may be found in [11]. Because the acoustic outputs are probabilistic, retraining an HMM to the changing habits of one speaker, or to entirely new speakers, is feasible.

HMMs may be cascaded. That is, the output of one set of models can serve as the input to a different set of models. For example, triphones may be determined by an HMM operating on the acoustic level. The triphones themselves, with probabilities attached, can be fed into a higher level HMM to determine the most likely word. At this level phonotactic and other constraints are taken into account. These results could be cascaded into a yet higher level HMM for phrases or sentences that incorporate the syntactic rules of the language.

3.9 Future Challenges I

HMMs own the present. Nearly all successful commercial systems are based on them. Lurking in the research laboratories of corporations and universities are the seeds for the basis of speech recognition in the coming century. These include auditory models, wavelets, and artificial neural nets. Of these, artificial neural nets are by far the most promising for speech recognition in the third millennium.

3.9.1 Artificial Neural Nets (ANNs)

All of speech recognition boils down to pattern recognition. Various acoustic patterns are associated with various speech sounds. The difficulty is overlapping patterns. Sometimes an acoustic pattern that is usually associated with a [b] is produced by a person saying a [p]. HMMs attempt to deal with this problem on a statistical, probabilistic basis. Artificial neural nets attempt to deal with the problem from a classification point of view: When given a certain acoustic pattern, they classify it into one of a hundred different categories, each of which is associated with an allophone of the language. This is just another way of approaching the underlying pattern recognition nature of speech recognition.

One attraction of ANNs—often called simply **neural nets**—is that they purport to carry out computations in a way similar to the human brain. They have been used in many nonspeech areas such the computer recognition of printed characters or of human faces. Their use in speech recognition is incipient, with most extant systems operating on the research level. (The company of Lernout & Hauspie is reported to use neural network technology commercially [12].)

ANNs consist of numerous processing units, any one of which is capable only of simple arithmetic operations such as addition and comparison, or simple logical operations such as *or*-ing and *and*-ing, all wired together in a complex network that resembles the contents of that kitchen drawer into which you throw all your rubber bands. At one end of the tangle are the inputs, at the other end the outputs, and in between these endpoints are one or more so-called hidden layers. (Speech scientists are secretive types—they're always hiding something.) A generic ANN might have the appearance shown in Figure 3.12.

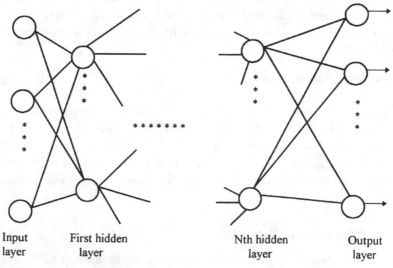

Input layer	First hidden layer	Nth hidden layer	Output layer

Figure 3.12 A generic artificial neural net.

Each input value is fed to several of the next processing units. These, in turn, send their outputs to the next layer's processing units, and so on. Each processing unit of the final, output layer will compute to (near) zero except the one determined to be the correct classification, which computes to (near) one. The point of the net is to classify the various combinations of inputs into one of the outputs. For example, in performing letter recognition there would be 26 outputs, one for each letter. If the pattern to be classified is a four-by-four grid containing ones and zeros according to whether a line of the letter passes through it (one) or misses it (zero), then there would be 2^{16} inputs. See Figure 3.13 for illustration.

Like HMMs, ANNs require training. By exposing an ANN to certain inputs and telling it what the output should be, the computing apparatus of the hidden layers can accommodate itself to make that classification. The accommodation takes place in small increments so an ANN that performs complex classifications may require a large amount of training.

The computations of the individual processing units—dare we call them "neurons"?—are relatively simple, though in collaboration they can achieve complex results. Let's look at a simple speech recognition ANN that takes as input the first and second formants of some person's pronunciation, and attempts to classify it into one of the four front vowels [ē, i, e, ă].

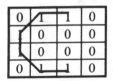

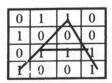

Figure 3.13 Four-by-four grids containing the letters 'C' and 'A.' If a part of the letter is in a grid square, it is given the value one, otherwise zero. The patterns of zeroes and ones can provide the input to an ANN which attempts to classify them into one of the 26 letters of the alphabet.

Logically we'll design the ANN so that initially it performs the following arithmetic/logical operations:

- If F1 is between 200 and 400, and F2 is between 2,000 and 2,400, then the output is [ē].
- If F1 is between 300 and 500, and F2 is between 1,900 and 2300, then the output is [i].
- If F1 is between 400 and 600, and F2 is between 1,800 and 2200, then the output is [e].
- If F1 is between 500 and 700, and F2 is between 1,700 and 2100, then the output is [ă].

The numbers are all in Hz. Figure 3.14 schematizes the process.

The astute reader immediately notices a conflict. What if F1 is 320 and F2 is 2,100? There is some doubt as to whether the output will be [ē] or [i]. That's okay at the start because we want to encompass all possibilities, and initially we allow the ANN to make arbitrary decisions. Training the ANN to a

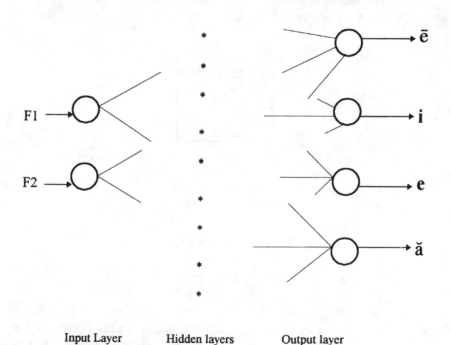

Input Layer   Hidden layers   Output layer

Figure 3.14 An ANN for classifying an input of the first two formants into one of the four front vowels [ē, i, e, ă].

particular person will reduce the ranges. If that person is an adult male, the actual formants will be in the low end of the ranges, whereas a female child's formants would be in the high end. Suppose we are training an adult male voice. The mysterious hidden layers would, after a number of training passes, adjust the ranges to be more in line with the actual voice, perhaps something like this:

- If F1 is between 200 and 300, and F2 is between 2,000 and 2,100, then the output is [ē].
- If F1 is between 300 and 400, and F2 is between 1,900 and 2000, then the output is [i].
- If F1 is between 400 and 500, and F2 is between 1,800 and 1900, then the output is [e].
- If F1 is between 500 and 600, and F2 is between 1,700 and 1800, then the output is [ă].

Now the ANN is well-trained to give more or less unequivocal results. Obviously, the numbers have been idealized, but the general idea is valid.

A real speech recognition ANN would take a far richer set of acoustic information as input than merely F1 and F2, and its hidden layers would be complex. Its output would be any of the kinds of speech units we have discussed. Schematically, it might look something like Figure 3.15.

Rabiner and Juang observe four strengths of ANNs that may make them the outstanding candidate for computer speech recognition in the next century [9]. First, they lend themselves to parallel computation. Parallel processing is a way of overcoming the speed limitations of a single processor. ANNs are robust and noise-tolerant due to the highly distributed nature of their computations. ANNs are readily adapted to small changes in input and are, therefore, suitable for speaker-independent recognition. Finally, ANNs have the mathematical properties to replicate an arbitrarily complex system of classifications.

3.10 Errors

Errors in speech recognition may have drastic consequences. If Luke Skywalker's command to his spaceship's computer is "deploy the defensive shields," and it is misrecognized as "destroy the defensive shields," our hero might not live to earn his producers another hundred million dollars. There can be little doubt that human mistakes in speech recognition have contributed to military disasters, political fiascoes, lovers' quarrels, and broken friendships.

There is little we can do about human error, but we can address the episodes that lead people to say "to err is human, to really foul things up you

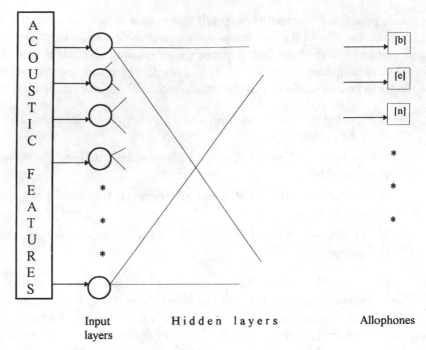

Figure 3.15 An ANN for speech recognition.

need a computer." The aim of the rest of this chapter is to identify the types of errors that may occur in computer speech recognition, trace them to their source, measure them, detect them, and ultimately correct them.

3.10.1 Types of Errors

When humans misrecognize speech they are capable of doing so on many different levels. They might miss a sound, mistaking *train* for *strain;* they might interpret syntax wrongly, mistaking *no smoking section available* to mean "there's no place to smoke" rather than "there is a place where smoking is prohibited"; or they may make incorrect semantic interpretations, taking "I can't recommend that math course too highly" to mean that the course was lousy rather than great.

Errors can only be made in realms where processing takes place. For humans recognizing speech, those realms are replete throughout all levels of the language from phonetic to semantic. For computers recognizing speech, the errors are confined to the kinds of speech units on which recognition is based.

If the speech units are words, only word errors can occur. If there are reference models for the digits zero through nine, it would be impossible for the recognizer to mistake the utterance *nine* for *fine,* even though the *f* "is there" in *five.* It may confuse an utterance of *nine* for *five* or *one* or any other digit, or deny recognition altogether, but the possibilities are limited to those cases. On the other hand a recognizer may substitute an [f] for an [n] if its speech units are allophones.

3.10.1.1 Errors in Discrete Speech Recognition

No human wants to speak discretely except when saying, with teeth clenched, *leave..my..thin..mint..Girl..Scout..cookies..alone.* So why do most speech recognizers require discretely spoken speech? The possibility of error is limited and controllable with discrete speech, whereas it may easily range out of control with continuous speech. The simple phrase *recognize speech,* when spoken without a break between words, can be misrecognized in some surprising ways, as we saw illustrated in Figure 3.9: *wreck a nice peach, wreck a nice beach, wreck an ice peach, wreck an ice beach, reckon ice peach, wreckin' eyes peach,* and on and on. By knowing in advance that there are exactly two words, and knowing where the break between them is, most of the misrecognitions won't be made.

The first error type we'll look at is one that isn't regarded as an error at all when it occurs among human speakers, and that's a failure to speak loud enough. We all know someone who, out of perversity, social insensibility, or a diminution of the lungs, always speaks one notch below your threshold of hearing, forcing you to read lips, misinterpret what is said, or constantly respond "eh? eh?" A computer speech recognizer also needs to receive signals that exceed a certain intensity, otherwise it rejects the utterance out of hand, a **deletion error.** It may ignore the input altogether or treat it as background noise, possibly providing negative feedback in the form of a screen icon or a buzz.

It would be nice if you could adjust your ears to hear sound at a very quiet level, but that's not possible. Speech recognizers can be made extremely sensitive to sound but there is a disadvantage: noise. If the recognizer was tuned to accept very softly spoken speech as input, it would also be that much more sensitive to environmental noise and **insertion errors** would be more common. That's when the dog barks at the trash collector and the recognizer thinks you said "delete all my files." Thus, the recognizer's sensitivity, or **gain,** is set at a level that requires you to speak in a normal voice as if addressing another person several feet away. This prevents interference from all but very loud background noise.

Once the recognizer has accepted your spoken input, only one of three outcomes is possible. The input is accepted, and the correct speech unit is identified. The input is accepted and an incorrect speech unit, called a **substitution error,** is identified. The input is **rejected,** which means that the recognizer was unable to choose a speech unit that it found correct.

A rejection may or may not be an error. If the recognizer only has reference templates for the digits, and the user says *sheepdog,* rejection is the proper response because that word is not in the vocabulary. Often, though, the user speaks a digit and for any of several reasons the recognizer does not perceive it as a vocabulary item and rejects it—a true error. **Rejection errors** are an annoyance. In practice, you say a word and you get some kind of negative feedback that means you have to repeat the word. Irritated, your subsequent pronunciations of the word are likely to be even less similar to the training template than the one that was rejected in the first place, and other rejections follow.

Substitution errors are insidious if not detected. The recognizer thinks it's right, but it's not. This may lead to erroneous data input, spurious transactions, incorrect commands, and other kinds of trouble. Table 3.3 summarizes the error situation for recognizers that accept discrete speech units.

3.10.1.2 Errors in Word Spotting

Word-spotting errors are similar to errors in discrete speech recognition, even though word spotting recognizers accept continuous speech input. If a word in the input stream is missed, the error is a **false reject.** If a word is incorrectly recognized, it is a **false alarm.**

3.10.1.3 Errors in Connected/Continuous Speech Recognition

The potential for errors in recognizing nondiscrete speech is tremendous. Low-gain and rejection errors, in which the recognizer digs in and refuses to perform, occur here as in discrete speech. Both insertion and substitution errors show up as in discrete speech. Deletion errors occur not only because of

Table 3.3
Errors in Discrete Speech Recognition

Type of Error	Description of Error
Too low gain (deletion error)	The input speech was spoken too quietly to register; it may be ignored by the recognizer and treated like background noise
Insertion error	Nonvocabulary input or noise is recognized as a speech unit. Noise may come from the user as throat-clearing, coughing, etc. or from the background such as a door slam
Rejection error	Valid input (an utterance in the vocabulary) is accepted from the speaker but the recognizer is unable to determine what was said
Substitution error	Valid input is accepted and misrecognized as a different speech unit than was spoken

too low a gain, but because in continuous speech the recognizer may overlook some input. There is also the possibility that one speech unit will be taken for two or more speech units, or, in general, that N speech units will be taken for $N + I$ speech units. For example an input of *recognize* could be mistaken for *wreck a nice* ($N = 1$, $I = 2$). Or *reckon ice* might be taken for *lick a nice*. ($N = 2$, $I = 1$). Such errors are termed **splits**.

The other side of this nasty coin is that two or more speech units will be recognized as one unit, or, in general, that N speech units will be taken for $N - I$ speech units. For example, an input of *the row* is mistaken for *zero* ($N = 2$, $I = 1$). Such errors are **fusions**. Neither splits nor fusions are possible in discrete speech recognition.

With discrete speech the errors arrive one at a time. One word is deleted or inserted at a time; one word is substituted for one word. However, when speaking continuously in sentences, multiple errors of the same type are possible. More than one word may be deleted from an input string, and in a noisy environment, more than one word may be inserted into an utterance.

Some years ago my university laboratory staff tested several continuous speech recognizers using a military vocabulary and issuing fairly short, battle-field-oriented sentences. A compendium of the errors we encountered is found in Table 3.4.

Some of the output was hard to classify. What should we say about an input of *co-ax fire on target* that is recognized as *co-ax fast?* Is it a substitution of *fast* for *fire* plus the deletion of *on target,* or is it a fusion of *fire on* into *fast* plus the deletion of *target,* or is it a fusion of *fire on target* into *fast?* An even more difficult instance to classify was an input of *co-ax on target* recognized as *gunner can't load sabot fast.*

The error classifications of Table 3.4 represent an attempt to "read the mind" of the recognizer. How do we really know that the recognition of an input of *tank steady right as tank tank steady* is a split of *tank* into *tank tank* and a fusion of *steady right* into *steady,* rather than the insertion of *tank* at the beginning and the deletion of *right* at the end? We don't, actually, and we couldn't without recourse to the computer software of the recognizer.

These difficulties have led some analysts to reduce the error types to three: substitution, deletion, and insertion. All of the data could be accounted for, eschewing the attempt to describe the actual mechanism, which might truly be splits or fusions. For example, the recognition of *I can't go any faster gunner* as *right M-48 faster gunner* could be described as a substitution of *right* for *I,* the deletion of *can't go any* and the insertion of *M-48.* Unfortunately, this isn't the only possibility. What about the substitution of *right* for *I* and *M-48* for *can't* and the deletion of *go any?* Analyzing *M-48* as a fusion of *can't go any* is attractive because the phonetic similarity of the two expressions makes the analysis plausible.

Table 3.4
Some Errors in Continuous Speech Recognition

Error Type	Spoken	As Recognized
Simple substitution	I can't fire faster	Tank can't fire faster
Multiple substitution	Driver move tank out slower	Driver move tank fast load
Simple deletion	Co-ax fire on target	Co-ax fire target
Simple insertion	Co-ax fire on target	Co-ax fire on target go
Multiple insertions	Gunner cease fire	Gunner cease fire tank tank
Multiple insertions and deletions	M-60 turn rear	Can't go sagger M-60
Fusion	[Move tank] slower right	[Any] slower right
Split	[Gunner identified] target tank	[Sabot sabot I] target tank
Substitution and insertion	Forward steady steady steady stop	Forward can't steady steady tank stop
Deletion and substitution	Watch for sagger load ammo	Watch sagger load tank
Deletion and insertion	Driver move out	Move out tank
Substitution and fusion	I [can't go any] faster gunner	Right [M-48] faster gunner
Split and deletion	[Ammo] out on M-60 tank	[Can't go] out M-60 tank
Split and fusion	[Tank] {steady right}	[Tank tank] {steady}

Anyway, who cares? What difference does it make, if we know the underlying mechanism of errors? There are two concerns. First, if we wish to measure and compare the performance of various recognizers, we need a precise method of error measurement. Second, if we understand where errors actually come from, as opposed to merely tabulating them, we may be able to find better ways of avoiding them. Both these topics will be taken up in Section 3.11.

3.10.2 Error Tolerances

You've been mugged. You got a good look at the mugger, and now you're at the police station reviewing mug shots of known criminals in your neighborhood. What are the criteria for identification such that a warrant will be issued? First, the mug shot must resemble the mugger to a fairly high degree, though not perfectly since people change but photographs don't. Second, if a person is to be accused, his mug shot must resemble the mugger to a sufficiently higher degree than all the other mug shots in the police files.

There are two criteria, then. The absolute one of resemblance, called the **alpha tolerance,** and the relative one in which one individual's resemblance is enough stronger than that of all the other candidates, called **delta tolerance.**

The same principle is in effect during speech recognition. To be recognized, an input speech unit must resemble at least one reference model to a degree determined by the alpha tolerance, and among those resembled, it must stand out to a degree determined by the delta tolerance. Figure 3.16 illustrates.

In Figure 3.16, the distances d1, d4, and d5 are all less than alpha, so reference models 1, 2, and 5 are all candidates for the unknown speech unit, with 1 as the leading candidate. Since the differences d4 − d1 and d5 − d1 are both greater than delta, reference model 1 is recognized as the input.

If no reference model is within the alpha tolerance of the input unit, the recognizer rejects the input. If the input was not in the vocabulary of the recognizer, the rejection is justified. For example, it might be a dog barking, a door slamming, or the user coughing. If the input is in the vocabulary, then a rejection error has occurred. If several reference models are within the alpha tolerance, but no one of them stands out from among the others, rejection also occurs.

The two tolerances may be adjusted. For example, an excessive number of rejection errors may indicate too stringent an alpha tolerance, and it may

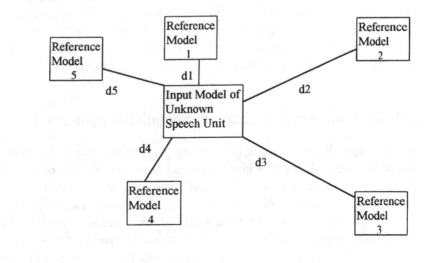

Figure 3.16 An illustration of alpha and delta tolerances.

be relaxed. The downside is an increased likelihood of accepting false input. Excessive rejection errors may also come from too stringent a delta tolerance, which may also be relaxed, with the downside of an increase in substitution errors.

Substitution errors may often be reduced by tightening the delta tolerance, that is, demanding that one reference model stand well apart from all other candidates as a condition of acceptance. The downside is an increase in rejection errors.

The tolerances may be adjusted for optimal behavior, where "optimal" is relative both to the user and the application. Certain applications may be such that substitution errors could lead to grave consequences, whereas rejection errors are merely inconvenient. These would have large delta tolerances. In others, substitution errors may be of little consequence, say in the presence of robust error correction (discussed in Section 3.13). In such cases, the delta tolerance may be lowered to reduce the frequency of rejection errors, which has the effect of improving throughput.

The alpha and delta tolerances work well for discrete recognizers when the speech unit is a word or phrase, and vocabularies are relatively small and well-chosen. With large vocabularies it may not be possible to adjust tolerances so that a unique reference model is always found. In such cases, a small group of reference models may stand out from the rest, and some further processing takes place to narrow the choice to one. This also occurs with continuous speech based on smaller speech units such as allophones. For example, in the context _ack, b, d, and g may meet both the alpha and delta criteria, but the system determines that only back is an actual word, and chooses b.

3.11 Performance Evaluation of Speech Recognizers

Buying a speech recognizer is like buying a car. There are many models to choose from, and they vary widely in price. (Tetschner lists 75 companies that sell speech recognizers [13].) Moreover, you must match your needs to the model. If you're single, cool, and move with a fast crowd, a dashing sports car is your cup of tea. If you're married with children a station wagon might suit you. If you take your muddy dogs everywhere it might be best to have a pickup.

Similarly, you choose whether to purchase a particular type of recognizer based on the application you intend to use it for. That will dictate whether it's to be speaker-dependent or -independent, has a large or small vocabulary capacity, and accepts discrete or non-discrete speech input.

Before purchasing a vehicle you can get some idea as to how it performs by looking in one of the magazines that publishes test results such as acceleration rates, gas mileage, and maneuverability. The demand for speech recognition systems is not yet high enough to support a glossy magazine that might be called "Recognition Trends" or "Recognizer and Speaker," in which testing is regularly reported, but at least some models of large vocabulary recognizers are benchmark tested by ARPA, and by the *National Institute of Science and Technology* (NIST), formerly the National Bureau of Standards.

Only similar kinds of recognizers are compared: You wouldn't evaluate a Rolls Royce the same way you evaluate a Chevette, and similarly, it makes little sense to compare a discrete speech recognizer with word-level reference models against a continuous speech recognizer whose reference models are based on triphones. Testing is carried out on prerecorded standardized databases[3] and follows a rigid protocol, just as automobiles are tested on carefully designed tracks by trained drivers. A number of criteria may be used in the overall assessment of a speech recognizer but by far the most important is robustness, or immunity to error.

3.11.1 Error Rates

In testing for a particular characteristic it's important to hold all other variables as constant as possible. If you're measuring the acceleration rates of automobiles, each run should carry a driver of the same weight, take place on the same track under similar wind conditions, and so on. In testing and comparing speech recognizers for robustness, the environmental noise level should be the same, the same communication channels should be used including the input microphone, and all controllable parameter values such as the alpha and delta tolerances should be set to the same levels. The creation of the reference models should be based on the same prerecorded data, and the same training procedures should be followed for each machine to be tested. The stage where actual recognition is measured should be based on identical protocols using the same set of prerecorded utterances in every case.

3.11.1.1 Discrete Speech Recognizers

For discrete speech recognizers, a typical testing protocol is to provide input in the form of vocabulary words with a smattering of nonvocabulary words. The three error types, insertion, rejection, and substitution, are tabulated and reported as a percentage. Gain is adjusted, so deletion errors are not an issue.

---

3. Many of these databases are available from the Linguistic Data Consortium (LDC) at the University of Pennsylvania, 441 Williams Hall, Philadelphia, PA, 19104-6305; telephone: (215) 898-0464; and e-mail: ldc@unagi.cis.upenn.edu

(Gain sensitivity may be tested separately.) Table 3.5 describes the measurements that are necessary.

In ascribing a single figure of merit to a recognizer, either total recognition accuracy or total error rate is reported. The insertion error rate is often not reported and not measured, partly justified by the fact that users can (and perhaps should) learn to confine themselves to the trained vocabulary. (Immunity to background noise is another issue. We'll discuss it in Section 3.11.2.1.) The fraction of substitution errors is useful since these are potentially more damaging than rejection errors. This number may also reflect the alpha and delta threshold settings. It may be useful to repeat testing with these settings at various values.

A recognition accuracy of 0.98 doesn't appear significantly worse than 0.99, but the former reflects an error rate double the latter—0.02 versus 0.01. In terms of error rate, an accuracy of 0.90 is 10 times worse than an accuracy of 0.99, not merely 10 percent worse. Recognition accuracy figures are misleading this way, and it's always best to be aware of the error rate.

3.11.1.2   Word-Spotting Recognizers

Error rate measurements for word-spotting recognizers is similar to that of discrete speech. Both false rejects and false alarms are tabulated and reported as a percentage of total input. Total input could be measured in terms of total words or in terms of seconds of time. Since word spotting recognizers accept **all** continuous speech, it may not be feasible to measure it in terms of words.

3.11.1.3   Continuous Speech Recognizers

Measuring error rates in continuous speech is more difficult than on discrete speech, as suggested by the data in Table 3.4. It doesn't have to be. One could simply report the percentage of input strings that are error-free as a measure of recognizer accuracy. This would be the strictest possible measure, though the results may be misleading. The method fails to discriminate between a recognizer that makes 10 errors per input string on half its input, and one that makes two errors per input string on half its input. Both recognizers would have an error of 0.5, though one is clearly doing a better job than the other. In particular, two errors per input may be much easier to interpret or correct with further processing, whereas 10 errors per input is likely to produce unusable results.

Another method is simply to record the fraction of individual words correctly recognized. The overall error rate is calculated by subtracting that number from 1.0. This method has the advantage of being easily automated and the disadvantage of being uninformative as to the kinds of errors that are occurring.

Table 3.5
Error Measurement for Discrete Speech Recognizers

Measurement (Range)	How Measurement Is Calculated
Total recognition accuracy (0–1.0)	Number of correctly recognized words ÷ Total number of vocabulary words spoken
Substitution error rate (0–1.0)	Number of substitution errors ÷ Total number of vocabulary words spoken
Rejection error rate (0–1.0)	Number of rejections of vocabulary words ÷ Total number of vocabulary words spoken
Insertion error rate (0–1.0)	Number of recognized (unrejected) nonvocabulary words ÷ Total number of nonvocabulary words spoken
Total error rate vocabulary words (0–1.0)	Number of substitution errors plus number of rejection errors ÷ Total number of vocabulary words spoken also 1.0 − total recognition accuracy
Fraction of substitution errors (0–1.0)	Number of substitution errors ÷ Number of substitution errors plus number of rejection errors

If insertions, deletions, substitutions, splits, and fusions are the error types, then error rates can be calculated by counting the number of errors of each type and dividing by the total number of words spoken, similar to some entries in Table 3.5. If 1,000 words are spoken and we count 50 insertion errors then the insertion error rate is 0.05, and so on for the other error types. This information could be combined with the fraction of input strings totally error-free, and the fraction of individual words correctly recognized, to give a comprehensive profile of a recognizer's error behavior. This is a difficult method to automate, because it is not always clear when an error is a simple split and when it is a compound error of insertion plus substitution.

Limiting the taxonomy of errors to insertions, deletions, and substitutions makes error tabulation easier to automate. Splits can always be interpreted as one or more insertion errors plus a substitution error. Fusions can be interpreted as one or more deletion errors plus a substitution error. As a consequence, however, the error of recognizing *zero* as *the row*—a simple split and arguably one error—would count as two errors, the insertion of *the* and the substitution of *row* for *zero*. Such an error (or errors) would weigh more heavily than the single substitution error of *Amazon* for *Mississippi*. This may or may not be justified. On one hand, mistaking *zero* for *the row* not only produces incorrect words, but incorrect word boundaries, a grievousness that may deserve an additional penalty. On the other hand, mistaking *zero* for *the row* is a natural kind of error, one that a person might make, and is therefore predictable, whereas the substitution of *Amazon* for *Mississippi* seems capricious and unpredictable. Being able to anticipate potential errors makes them easier to compensate for with further processing or with simple vocabulary changes, say *cipher* in place of *zero*.

Two colleagues and I proposed a somewhat different method of evaluation in [14]. Each input string tested would be awarded three numbers. The first would be the number of words correctly recognized divided by the total number of input words; the second would be the number of words "recognized" but not in the input, divided by the total number of input words; the third would be calculated in the following way. Let L be the number of internal word boundaries in the input string and L' be the number of internal word boundaries actually recognized. If $L = L'$, assign 1. If $L > L'$ assign L'/L. If $L' > L$ assign L/L'. This is a measure of how well the recognizer does on detecting word boundaries within the utterance. (The beginning and end of the string are technically word boundaries but are not counted.)

Let's look at an example. Believe it or not, one virtue of the method is that it is easily computerized because it is so straightforward. It just seems complicated. Consider the first entry in Table 3.4, where the input string is *I can't fire faster,* and the output string is *Tank can't fire faster.* The triplet of numbers assigned to this datum would be $\langle .75, .25, 1 \rangle$ because three out of four words were correctly recognized, one out of four was incorrect, and the word boundaries were all correctly detected. Table 3.6 is Table 3.4 repeated with the error triplets calculated.

In evaluating and comparing the performance of recognizers, the values could be averaged over a large number of input sentences and a triplet assigned to each recognizer. The first member of the triplet has a range in [0,1]. The best score is 1. It's a measure of how well the recognizer does in preserving the correct input words. The second number of the triplet is in [0, ∞]. The best score is 0, which indicates no insertions and no substitutions. Since insertions are theoretically unlimited, this number has no upper bound. However a value as high as 1 indicates as many wrong output words as input words, which should rate low. The third number is in [0,1] with 1 the best score. It is in part a measure of how well word boundaries are detected. It is also an indication of errors other than substitution errors, because insertions, deletions, fusions, and splits all affect the number of word boundaries.

A perfect score would be ⟨1,0,1⟩. Recognizers could be rated according to their Euclidean distance from this point in three-dimensional space, though other distance measures are possible. Indeed, since the triplet is derived statistically, several statistical methods for ranking recognizers are possible.

Ideally, identical testing procedures are used on all recognizers to be compared. Otherwise, the measurements have little significance. This is the approach taken by NIST and other neutral observers interested in valid assessments. Laboratories and manufacturers who test their own products (or competitor's products) are **always** suspect: This is an area where petty prevarication is widespread.

3.11.1.4 Factors Affecting Test Results

Give me any speech recognizer and I'll test it with guaranteed 100 percent accuracy on 1,000 test utterances randomly chosen from its vocabulary . . . of one word!

The size and similarity of the vocabulary of speech units affects the error rate. Large vocabularies are more likely to have similarly sounding words that may be substituted for one another. Even a small vocabulary with many similar sounding units suffers from the same defect. For example a vocabulary consisting of the names of the 26 letters of the alphabet is likely to engender many errors because of the sets of similar sounding words: {bee, cee, dee, ee, gee, pee, tee, vee, zee}, {a, jay, kay}, {I, why}, {en, em}. That is why a special "phonetic alphabet" such as the one used in radio communications is recommended for the input of letters: *alpha* for *a; bravo* for *b;* and so on. *Bravo* and *delta* are much less likely to be confused than *bee* and *dee.*

Standardized testing uses a fixed vocabulary so what's fair for one is fair for all. However, it is easy enough to prejudice the testing of an individual recognizer

Table 3.6
Some Errors in Continuous Speech Recognition With Scoring Triplets

Error Type	Spoken	As Recognized	Error Triplet
Simple substitution	I can't fire faster	Tank can't fire faster	⟨.75, .25, 1⟩
Multiple substitution	Driver move tank out slower	Driver move tank fast load	⟨.6, .4, 1⟩
Simple deletion	Co-ax fire on target	Co-ax fire target	⟨.75, 0, .67⟩
Simple insertion	Co-ax fire on target	Co-ax fire on target go	⟨1, .25, .75⟩
Multiple insertions	Gunner cease fire	Gunner cease fire tank tank	⟨1, .67, .5⟩
Multiple insertions and deletions	M-60 turn rear	Can't go sagger M-60	⟨.33, 1.0, .67⟩
Fusion	[Move tank] slower right	[Any] slower right	⟨.5, .25, .67⟩
Split	[Gunner identified] target tank	[Sabot sabot I] target tank	⟨.5, .75, .75⟩
Substitution and insertion	Forward steady steady steady stop	Forward can't steady steady tank stop	⟨.8, .4, .8⟩
Deletion and substitution	Watch for sagger load ammo	Watch sagger load tank	⟨.6, .2, .75⟩
Deletion and insertion	Driver move out	Move out tank	⟨.67, .33, 1⟩
Substitution and fusion	I [can't go any] faster gunner	Right [M-48] faster gunner	⟨.33, .33, .6⟩
Split and deletion	[Ammo] out on M-60 tank	[Can't go] out M-60 tank	⟨.6, .4, 1⟩
Split and fusion	[Tank] {steady right}	[Tank tank] {steady}	⟨.67, .33, 1⟩

by choosing an "easy" vocabulary, that is, one in which all the units are phonetically very distinct.

Microphone quality affects recognizer performance, especially if different microphones are used in the training and testing stages. Even the placement of the microphone may be important. Time and time again, I've seen recognition performance deteriorate because the microphone shifts position during usage.

The channel through which the recognizer receives input affects its performance. It would be ineffective to compare recognizers where one gets its input through phone lines while the other gets its input directly. If a recognizer operates on telephone input, its reference models should either be created over phone lines, or prefiltered in such a way as to replicate the effect of phone line transmission.

Background noise degrades recognizer performance. Under controlled testing conditions, noise is often eliminated. For recognizers designed to work in a noisy environment, noise must be simulated by the testers. This can be done electronically, or done the way I did it once. I was asked to evaluate some speech recognizers in noise. I took a high-quality portable tape recorder and recorded the everyday sounds of a tire shop, an automatic car wash, a sports bar, and a disco club (it was a while back). I mixed the sounds up to create a cacophony on tape which I played at the same volume level through identically placed speakers each time I tested a recognizer.

If the utterances used for training and testing are not prerecorded, unacceptable variability is introduced. Human beings are not capable of behaving identically from event to event, whether asked to hit a home run in a baseball game, or repeat the word *anaesthetist* consistently. However, some people are better than others in their consistency of pronunciation. (They're called **sheep** in the trade.) When such a person trains reference models, then uses a recognizer, the results may be quite good. Other people (**goats**) are unable to pronounce consistently. When they use the same speech recognizer the results may be quite poor, even though other factors are held constant. However, even sheep will produce poor testing scores if they train reference templates while healthy but use the recognizer during an illness, when under stress, when intoxicated, and so on.

For testing speaker-independent recognizers the same cautions regarding human testers apply. To achieve valid results, prerecorded standardized utterance databases should be used. They should reflect the variety in the population where the recognizer is to be active. If speaker-independent speech recognizers are to be used in long-distance directory assistance, they must be able to recognize vocabulary words in the major American dialects, for example both [faiv] and [faːv] for "5." Thus, they should be evaluated using both these pronunciations.

With all these factors one can easily see why no manufacturer of speech recognizers ever had to admit to recognition accuracy below 0.99. Even without one word vocabularies, conditions could be made so ideal—clear channels, noise free, use of sheep—that only the most bumblingly designed recognizer would fail to achieve high recognition rates.

My own experience over the years has made me very wary of manufacturers' claims for their own products. (Is anybody not?) I realized that recognition rates declined in a predictable way, summarized in Table 3.7.

3.11.2 Other Factors

Recognition accuracy is the most important factor in the performance evaluation of speech recognizers, but it is not the only factor.

3.11.2.1 Noise Immunity

A high recognition in a quiet laboratory may turn into a low recognition rate in a noisy office. In assessing recognizers it is necessary to know about their sensitivity to noise.

There are three sources of noise in speech recognition: the speaker, the background, and the channel(s). You can get a sense of the dimension of speaker noise by following a two-year-old child around for a day. Don't skip mealtimes, where some of the best sounds are to be heard. You will hear a panoply of acoustic effects including lip smacks, coughing, and weird vocalizations. All these sounds also emanate from adults using speech recognizers. Sophisticated recognizers can be designed to detect and ignore nonspeech sounds at the start of an utterance.

Table 3.7
The Decline of Performance as Recognizers Emigrate From Manufacturer's Lab Through an Academic Lab and Out Into the Real World

Testing Environment	Testing Results (Recognition Rates)
Manufacturers laboratory using their own personnel under ideal conditions	99+% correct
My laboratory using my own personnel for testing	2–5 percentage points worse than previous
My laboratory with outside subjects brought in to test the system	2–5 percentage points worse than previous
My laboratory with outside subjects brought in to use the system for some pseudo-application	2–5 percentage points worse than previous
User's environment, actual applications, conditions not as closely controlled as in the laboratories	2–50 percentage points worse than previous

Mid-utterance dysfluencies are more difficult to handle. They include not only the common, revolting sounds of snuffling and gagging, but also hesitations, stuttering, and attempts to correct mistakes in the middle of utterances ("Delete docu . . ., I mean save document, aargghhhh!"). Modeling such behavior so it can be detected by the speech recognizer and accounted for is currently a subject of research.

Measuring the immunity of a speech recognizer to speaker-induced noise is no more complicated than having prerecorded test utterances with such noise purposely introduced. Performance on those particular sentences can be singled out as an indication of noise robustness.

Background noise includes any sounds not from the user that impinge on the microphone of the recognizer. Such environmental noise may cover the entire audible spectrum at all decibel levels. The noise may range from a steady, unchanging sound such as the hum of a motor, to widely fluctuating sounds such as a door slamming, foot falls, vehicle backfires, sirens, etc. Determining immunity to background noise might be done by including such noise in the prerecorded test utterances, or testing the system while simulating the noisy background from a tape recording of various kinds of environmental noise.

The source of channel noise is either the input microphone or the telephone system if used. The performance of a recognizer may depend heavily on matching the microphone type to the application. Unidirectional microphones are generally used since input comes from a single location, namely the user's mouth. Noise-canceling microphones are needed in noisy environments, and in a small enclosed location such as a kiosk, where reverberation is strong, a close-talking microphone may give the best performance. Testing can take place in the usual way, repeated for various combinations of microphones and environments, to give performance statistics for each case.

If a recognizer is not designed for usage over telephone lines, it may perform poorly. The telephone system acts as a band-pass filter, eliminating or attenuating frequencies below 300 Hz and above 3,400 Hz. Recognizers that rely on frequencies outside this range would perform poorly on telephone data. If the training of reference models takes place directly through a microphone, but usage is over phone lines—or vice versa—the results may be disastrous.

Databases of utterances made over phone lines are available for assessing the effectiveness of speech recognizers intended to operate in that environment. These may be obtained from the Linguistic Data Consortium.

3.11.2.2 Ease of Use

When buying a new car no amount of published test results replaces a road test. Great performance statistics mean little if you find the car uncomfortable to sit in or hard to handle. Speech recognizers may be judged analogously.

A good speech recognizer will be effective on a vocabulary size suitable for the intended applications. If you stretch the limits of vocabulary capacity, error rates may rise.

Once operational, it is easier to add, delete, or change vocabulary words in some models than others. If your application has a lot of vocabulary changes, this factor may be important. Training procedures for reference models also vary. Some recognizers require multiple repetitions of vocabulary words, others just one or two repetitions. These features are comparable to an automatic transmission or power seat. They are a matter of convenience and their importance varies from usage to usage.

We'll discuss error correction in Section 3.13, but the ease of recovery from errors is a factor that goes into the performance evaluation of a speech recognizer. A well-designed interface for handling errors is a desideratum.

The user's age and whether he or she took music lessons may be an ease-of-use factor. We old dudes who were forced to take piano lessons and learned to type on an old Smith Corona in the eighth grade have never outgrown our need for the keyboard. Our kids, bless 'em, have little use for fingers other then the two they use on the left and right mouse buttons. A well-designed recognizer interface should give the user plenty of mouse versus keyboard options in carrying out the basic tasks of training, usage, error correction, vocabulary modification, and so on.

Speed! Recognizers must operate in near real time to be effective. In most cases this means no more than a one-third second delay between utterances for discrete speech, and no more than a few seconds delay after the completion of an utterance in continuous speech. A large vocabulary may cause processing delays since more reference models must be compared. Some speech recognizers allow a trade off of speed for accuracy. Rather than exhaust all processing options in a search for the highest confidence recognition, these systems allow the user to curtail processing in exchange for a possibly higher error rate. Depending on the application, different levels of performance can be set.

Finally, good documentation can make a difference. It should explain all the variables that affect recognition performance in a clear and concise way so that they can be exploited. This includes the alpha and delta tolerance settings, other special settings, optimum placement of the microphone, creation and modification of reference models, and adaptation to noise.

3.12 Error Reduction

Errors in speech recognition are like sins. You can work on eliminating them in the first place, or you can atone. In speech recognition it's best to reduce the number of errors in the first place. Atonement takes the form of error correction, which works if the error is mild. In this section, we discuss ways of reducing errors. In Section 3.13, we'll look at correcting errors once they're made.

3.12.1 Environmental Effects

Anything that can be done to make the noise during usage the same as the noise during the training of reference models will reduce errors. That means that if the reference models are trained in a quiet environment, the recognizer should be used in quiet. If you intend to use the recognizer in the cockpit of a helicopter, it should be trained there. It may not perform as well in a hubbub as in a quiet room, but it will do a lot better trained and used in the noisy environment rather than trained in one and used in the other.

It's obvious why noise, or noise differences, produce(s) errors. The situation is no different with machines than with humans. However, noise in speech recognition produces another, unexpected effect. In the presence of background noise, people naturally tend to speak louder. This is called the **Lombard effect**. When you speak louder the quality of your voice changes, hence the probability of matching up with the correct reference model goes down.

The environment should be free of air-borne irritants. As we noted, coughing, throat clearing, and snuffling all contribute to a rise in error rate. A well-situated microphone and a sturdy chair, when possible, will reduce user fatigue and the concomitant voice changes that result in errors. The chair should not squeak!

3.12.2 Human Factors

If you have a pet like a dog or a cat, chances are you've been trained by the animal. I'm not suggesting that Fido or Tabby are inhabited by Pod People from Pluto with nefarious designs on the human race. Rather, our behavior is shaped by our interactions. When the dog scratches the door, you let her out; when kitty rubs against your leg, you reach down and stroke him. You've been trained!

Just as users train speech recognizers to their voice, the recognizers train users to their own idiosyncrasies. You soon learn not to raise your voice at your speech recognizer, because when you do, it ignores you. Better not whisper either, else you'll be punished by the cold shoulder. An excellent way, perhaps the best way, to reduce errors in speech recognition is to train users.

Earlier we classified people as sheep or goats. Speech recognizers work better for sheep than for goats, in part because sheep have more consistent pronunciations. Sheep are also better learners, or put another way, are more easily trained. They pick up on the little tricks that make the speech recognizer perform better. Goats are, well, stubborn as goats. They refuse to be trained and are doomed to a life of recognition errors.

In these days of genetic engineering you shouldn't be surprised to know that we can turn goats into sheep. Cooperative users, even if they're born goats, can learn the little tricks of coaxing good performance out of a recognizer. Some of the tricks are voice training, learning to speak consistently at a fairly constant level of amplitude. Other tricks are consistent placement of the microphone (or telephone), absence of fidgeting, and concentration.

The recognizer provides the cues on which sheep training is based. A rejection error is generally greeted with an unpleasant sound like a buzz; a successful recognition brings forth a pleasant sounding chirp. As you experiment on ways to reduce buzzes and increase chirps, you're well on your way to sheepdom. The recognizer literally conditions the user into modulating her voice in such a way as to reduce errors.

Where possible, recognizers provide visual feedback. This allows the user to see which inputs cause confusion. For example, the recognizer may consistently confuse *five* for *nine*. A common countermeasure is to retrain the reference models for the two words. If that doesn't help, and it may not because the words are phonetically similar, the user may choose a different word to mean "five," for example Italian *cinque,* pronounced "chink way," which is unlikely to be confused with *nine*.

Telephone interactions are different. Feedback is necessarily audible, as is confirmation on the user's part, and the whole business can get annoying:
You: nine nine nine five five five one two one two;
Recognizer: Did you say nine nine five five five five one two one two;
You: No, that was nine nine NINE five five five one two one two;
Recognizer: Please repeat;
You: (screaming) I said nine nine NINE five five five . . . Oh, the hell with it.

Most telephone applications that are voice-only do not rely on touch tones, and have a very small, distinct vocabulary. Any input not rejected is accepted without confirmation and processed. In case of error, the user always has recourse to a human operator.

Some telephone applications use word spotting. The user says, "I want to make a long-distance call" and the recognizer spots "long-distance." Request for confirmation is then appropriate. The system answers, "If you wish to make a long distance call, please say 'yes,'" and so on. Again, there are a large number of possible ways to design such interactions. It is important that experts in human factors be consulted before any particular system is put in place.

When a person feels stress, his or her voice changes. Many recognition errors are due to stress, and ironically, it's the recognizer that brings the stress about. Nobody likes rejection. A corollary is: nobody likes rejection errors from their speech recognizer. Intended or not, when you have to repeat an utterance there is a likelihood of tension in your voice. Since your reference models were made when you were not tense, the chance of a mismatch increases. A second rejection error brings more tension, and a third, yet more. The same applies to other error types. After each error, the chance for error increases. This can lead to downward-spiraling performance that often ends with a disillusioned user and a speech recognizer in the attic.

Speech recognizer users must learn to talk like airline pilots. They are unflappable: "From the flight deck ladies and gentlemen. We just lost our third engine. We will be attempting an emergency landing in the Bering Sea about one hundred miles west of Nome. Weather there is cloudy, forty-five degrees Fahrenheit, winds from the northeast at fifteen knots. We expect to be on the water at twelve-nineteen Alaska time, a short five minutes from now. Thank you for flying with us." Part of sheepiness is a resistance to the stress engendered by recognition errors, and this, fortunately, can be learned by goats.

The creation of reference models or templates is where many future potential errors can be nipped in the bud. They should be made under conditions as identical to those of future usage as possible. There is a contradiction here. The reference models are special. The user tends to spruce up for them, be on their best linguistic behavior, kind of like a first date. That's wrong, however. What's best is to make the reference models as mundane as possible because you'll get tired of sprucing up every time you use the recognizer. This is not an easy thing to do.

Experience may help you create reference models, which may be redone at any time. I was once part of a natural language project that input discretely spoken sentences. Each word in any sentence had to be trained, of course, but they were trained in isolation. When you speak a word in isolation the intonation falls off at the end. Try it. Say: *"first . . . second . . . third,"* or make someone else try it who doesn't know what you're listening for. The effect is usually there. Our problem was that in using the words in a sentence, the intonation was sometimes different than it was in training. For example, in *"is he first?"* there is a tendency to have rising intonation on *first*. To solve this problem we made three reference models for each word: one with falling intonation, one with level intonation, and one with rising intonation. Our error rate improved dramatically.

An experienced user gains a knack for continuous speech. For one, it's usually best to resist running words together—easier said than done since it's natural to do so. That is, in speaking *one, two, three* you should enunciate the *t* between *one* and *two*, rather than allowing it to merge with the *n*, giving a pronunciation that sounds like "won new three." Making more frequent pauses than usual also helps since it establishes some word boundaries. Finally, to the extent it doesn't contradict the previous two points, it's best to speak in a natural manner.

In all of the discussion in this section I have been assuming that users **want** the speech recognizer to succeed. Unfortunately, this is not always true. When recognizers are introduced into environments where they are seen as a threat, users may become uncooperative. Just as users can learn to be sheep, they can easily learn to be goats, and there are more ways of becoming a speech recognition goat than there are objects that a goat will eat. I once consulted for a company with several speech recognizers on a noisy factory floor. Recognition errors were through the roof when the workers used the system, though when we tested it under similar circumstances it worked fine. A little sleuthing revealed that folks felt that the recognizers would result in job loss. (They were really there for quality control.) Once management assuaged those fears, the system (magically?) started working better, but still not as well as in tests. Since every entry was logged, including the ID of the user, I suggested that management offer a weekly prize to the group whose recognizer had the best recognition rates. They did so and performance improved to the point where it exceeded that of our own tests. I said it earlier, but it bears repeating. Take it as axiomatic:

Speech recognition will not work if the users don't want it to.

3.12.3 Subsetting

Consider a hypothetical continuous speech recognizer whose vocabulary is *the, old, man, men, walk, walks, sleep, sleeps* and whose input will be any grammatical sentence using these words. Technically, the input is a set of an unlimited number of sentences if we agree that there is no limit to the number of repetitions of *old* in *the old, old, old, . . . man walks.*[4]

Suppose the user says *the old men walk* and the recognizer makes a substitution error and displays *the old men walks*. Hmmm, that's frustrating. Why isn't it smart enough to know that plural subjects take the infinitive verb form?

Some recognizers are that smart, leastwise, they can fake it. Here's how it's done. Of the eight vocabulary words, only two may follow *men* in a grammatical sentence, namely *walk* and *sleep*. The recognizer can be programmed so that after *men* has been recognized the next word spoken need only be compared with two reference models, those of *walk* and *sleep*, which comprise the **active**

---

4. Clearly, no one repeats *old* one million times, not even to describe Methuselah, but that is a physical limitation. From the point of view of the rules of English syntax, such an adjective may be repeated indefinitely.

vocabulary at that point. In effect, the **branching factor**—the size of the active vocabulary—is reduced from eight to two for the word spoken after *men*. The process of reducing the active vocabulary at points of the input is called **subsetting** or **syntaxing**.

Without subsetting, the branching factor at any point is equal to the number of words in the vocabulary. The average branching factor is called **perplexity** and is one measure of how difficult the recognition task is: The larger the perplexity, the more difficult. Perplexity is reduced by subsetting. A precise definition of perplexity may be found in [9].

In the current example there are many opportunities for subsetting. The first input word has a branching factor of four because it can only be *the, old, man, men* (*man walks* is grammatical as a generic statement). If *the* is recognized, subsetting can reduce the branching factor to three: *man, men, old*. After *old* the branching factor is three, and so on. By encoding this information the perplexity of the recognition task can be reduced by well over 50 percent.

Error rates are related to perplexity. A high perplexity engenders a greater possibility for error than a low perplexity. Recognizers that permit subsetting are therefore more likely to experience a reduction in error rate, all things being equal, than recognizers that don't.

Subsetting may also be achieved by both semantic and contextual constraints. If *glass, shatter,* and *shatters* were added to the vocabulary, semantic constraints could be used to reduce active vocabularies based on the semantic (but not syntactic) ill-formedness of expressions like *the glass sleeps or the man shatters*. Add *birds, fish, swim,* and *fly,* and in a context of (normal) locomotion, active vocabularies could be limited to prevent output such as *fish walk, men fly,* or *birds swim*.

In assessing recognizers that permit subsetting, it is important to program them identically. In a more complex task than our simple example, there may be several ways of implementing the subsetting routines.

3.12.4 Vocabulary Selection

A sure way of reducing the potential for error is a well-chosen vocabulary. The principal criterion in designing a vocabulary is functionality. The input speech has to get the job done. Within that limitation there is often room to maneuver. One rule to follow is the exact opposite of what Mark Twain advises:

> An average English word is four letters and a half. By hard, honest labor I've dug all the large words out of my vocabulary and shaved it down till the average is three and a half. . . . I never write "metropolis" for seven cents, because I can get the same money for "city." I never write "policeman,"

because I can get the same price for "cop." . . . I never write "valetudinar-ian" at all, for not even hunger and wretchedness can humble me to the point where I will do a word like that for seven cents; I wouldn't do it for fifteen.

Mark Twain's Speeches, ed. by A. B. Paine, 1923

Speech recognizers love words like *valetudinarian.* It's much better than *sick man,* which is easily confused with *chicken, thicken, thick sand,* and so on. In short, avoid monosyllabic vocabulary words; go for the fifteen cents! If the users will say *affirmative* for "yes," *negative* for "no" and *canis familiaris* for "dog," the error rate will go down. Many recognizers, impious contraptions that they are, will confuse *dog* with *god,* but few will confuse *canis familiaris* with *deity.*

In an application in which letters of the alphabet are to be input, it is best to use some kind of a military-like alphabet in which the letters are represented by words. Thus *A22B33C* might be entered as "Abel two two Baker three three Charlie." Of course, the training of the reference models would incorporate this approach.

Once a vocabulary is established on the basis of functionality and poly-syllabicity (a great word for recognizers), it may then be scanned for sets of po-tentially confusing words, and if possible, altered to avoid the confusion. A vo-cabulary containing the words *big* and *pig* invites confusion. Change *big* to *large* and the confusion goes away, and in all likelihood *large* will work wherever *big* did formerly.

With cooperative, repeat users, many vocabulary adjustments are possible that wouldn't be feasible for one-time users. In radio transmission, because *five* and *nine* are confusable, the word *niner* is spoken instead of *nine.* This is stan-dard usage between pilots and air traffic controllers. The same convention can be adopted by regular users of a speech recognizer to avoid a similar confusion.

Finally, a vocabulary with built-in redundancy may help reduce errors. A vocabulary that not only has *zero,* but also *oh, cipher, null,* and *nil* offers flexi-bility to the user. If *three* is often substituted for *zero,* the user can try one of the synonyms, which will have the same significance to the recognizer, but not cre-ate the same confusion when pronounced.

3.13 Error Detection and Correction

You've done everything possible to reduce errors in speech recognition, but still they come, more persistent than army ants, and occasionally more pesky and destructive. It's time to retreat, regroup, and rethink. It ain't over 'til it's over.

3.13.1 Feedback Systems

The best error detection and correction is done by the user. Rejection errors and low-gain errors are immediately apparent, and the user must repeat the input while making every effort to speak at the proper level in a manner consistent with the reference models. In speaker-dependent systems that means attempting to speak the same way during usage as during training. For large vocabulary partially speaker-dependent systems—ones in which the reference models adapt to the user—it means speaking consistently on all occasions. In speaker-independent systems it means articulating clearly, but not abnormally so.

There are many, many ways of interacting with a recognizer after a non-rejection error is detected. We'll consider four cases: discrete speech or continuous speech and visual feedback or no visual feedback (usually over the telephone).

3.13.1.1 Discrete Speech, Visual Feedback

Large vocabulary systems that take dictation use a computer screen to display the recognized input. A keyboard and/or mouse may or may not be available. (Quadriplegics can't use these devices.) The user usually, but not always, sees the error immediately. If it's an insertion error, perhaps the result of noise, the deletion key gets rid of it, or a special word can be spoken with the same effect. In the case of a substitution error, if it's caught immediately the word can be deleted (by key or by speech) and the correct word repeated.

If an error is caught downstream, correction is more difficult. If it's one or two words back, successive deletions and repetitions will work. If it's several words back the offender may be highlighted and deleted and the correct word repeated. If only voice control is available, the user can order the cursor to backup to the error, delete, and repeat. Protocols for accomplishing this vary from recognizer to recognizer.

If several errors slip through it is often most expedient to delete the entire utterance and start over. Again, this can be done via voice command.

Sometimes the system itself realizes that an error is probable. This occurs when several reference models are equally matched by the input. In that case the several corresponding words are displayed and the user asked to select one. The choice may be made via voice or keyboard. If none of the recognizer's guesses are displayed, it's back to delete and repeat.

What happens if an error correction voice command is mistaken by the recognizer? This occurs rarely because the correction vocabulary is relatively small and the words (should be) designed to minimize confusion. There are several protocols for dealing with this "error within errors" scenario. In a continuous speech recognition system whose design I once participated in, we used a made up word "forgetityouidiot" which was distinct enough from the other words to be always recognized, and even if it weren't, it could simply be repeated until it was. Forgetityouidiot canceled all bets and returned the user to the beginning of the previous transaction.

3.13.1.2 Discrete Speech, No Visual Feedback

Wouldn't it be nice to be able to talk to your automobile cellular phone? Instead of reaching down to dial a number, and plowing into the rear of that stupid bus, you could activate a speech recognizer and voice in the number, never taking your eyes off the road. You would like some kind of feedback, however. You'd hate to call Albania by mistake. This involves voice synthesis (Chapter 4) and may come after each digit is entered, so you say *nine* and the system repeats what it thinks you said; then you say *one,* the system repeats, and so on. Or the feedback can come all at once at the end. Correction of an error can be keyed to a word such as *backup.* Suppose you're trying to dial 999-1234. Here's a possible scenario:

You: Cellular awake!
Phone: Okay;
You: Nine;
Phone: Nine;
You: Nine;
Phone: Nine;
You: Nine;
Phone: Five;
You: Backup;
Phone: Okay;
You: Nine;
Phone: Nine;
You: One;
Phone: Nine;
You: Backup;
Phone: Okay;
You: One;
Phone: One;
You: Two;
Phone: Two;
You: Three
Phone: Three;
You: Four;
Phone: Four;
You: Transmit;

Phone: Okay.

It's less tedious than it looks, and a whole lot less tedious than extracting the front end of your new BMW from the rear end of a bus.

The other scenario goes like this:

You: Cellular awake;

Phone: Okay;

You: Nine nine nine one two three four end-of-entry;

Phone: Nine nine five one two three four;

You: Backup;

Phone: Okay;

You: Nine nine nine one two three four end-of-entry;

Phone: Nine nine nine one two three four;

You: Transmit;

Phone: Okay.

This method suffers from the defect that you always have to start over, even if the error was the last of ten digits, but that defect has been in the landline phone system since its inception and people are used to it. (Cellular phones generally allow you to back up and correct one digit at a time.)

Another use of speech recognition is verbal communication with the phone company computer. Often, word spotting of discrete words is in effect. Confirmation of each step in the process is required, so if an error is made, it can (hypothetically) be detected (underlined words are in the word-spotting vocabulary).

You: I'd like, uh, to make a <u>collect call</u>, you know, where the other guy pays, ha ha.

Phone: Do you wish to make a long-distance call? Please answer *yes* or *no*.

You: Uh, what was that, oh, <u>yes</u>, please . . . <u>no</u>, I mean <u>no</u>.

Phone: Do you wish to make a long-distance call? Please answer *yes* or *no*.

You: <u>No</u>, imbecile!

Phone: Please state the service you wish.

You: Can I make a dadgum <u>collect call</u>?

Phone: Do you wish to make a collect call? Please answer *yes* or *no*.

You: <u>Yes</u> (sigh!).

Actually, transactions usually proceed more smoothly, especially after users have accumulated some experience. (The sighing, muttering, and cursing do little good.) You may wonder why bother with speech at all when the same information can be sent via touch tone. There are two reasons. One is that touch tone protocols are incredibly annoying: *dial one for long distance, two for collect calls, three for credit card calls, four to speak with an operator, and five if you want relish on your hotdog.* Second, many phone systems around the world lack touch tone service. In such locations it may be economically wiser to leapfrog touch tone technology and go straight to speech recognition, even for ordinary dialing.

3.13.1.3 Continuous Speech, Visual Feedback

Because continuous speech recognition has some lag time, some of the strategies mentioned above are not available. By the time the user detects an error he's likely to be well beyond that point of the input. The most common strategy is to repeat the entire input, with the various options of keyboard or verbal commands available.

Human-like error correction strategies for continuous speech are being worked on in research labs but are not yet found in commercial products. If you were dictating to a person who was displaying the results on a monitor and said: "I'm going to the University of Southern Albania to give a talk on how to recognize speech" and you saw *I'm going to the University of Southern Albania to give a talk on how to wreck a nice beach* you wouldn't repeat the whole sentence. You'd simply say "that was re-cog-nize speech, not wreck a nice beach," or something similar. Or if you gave a stream of digits like 1133557799 and the human typed 1133597799, rather than repeat the entire string you would say "that 9 in the middle should be a 5." The difficulties of such intelligent error correction are apparent, but they are no longer beyond the technology horizon.

3.13.1.4 Continuous Speech, Nonvisual Feedback

Some systems will accept continuous speech over the telephone, usually strings of digits. At the end of the input the system repeats back what it thinks you said and asks for confirmation. There isn't any opportunity to make small corrections. The entire transaction is repeated in case of error. Some of the remarks made in the previous paragraph apply here, too. So if I request that the system dial 999-555-1515 and it asks me to confirm 999-555-1519, I would rather say "make that last digit a 5" than repeat the whole string.

3.13.2 Higher Levels of Linguistic Knowledge

The ultimate goal in speech recognition is for the recognition system itself to detect and correct errors. That's what people do. We rarely hear everything said to us perfectly. We are continually applying our human intelligence and knowledge when we recognize speech. We test hypotheses—could he really mean "wreck a nice beach?"—and use our better judgment to reject them and entertain others, all of which goes on at lightning speed beneath the level of conscious reasoning. Even so we make errors, or fail to understand. This occurs in noisy environments or when we are distracted, tired, or bored. The one advantage machines have over us is they do not get distracted, tired, or bored. If we could impart human intelligence and knowledge to them, they may in the long run turn out to be better speech recognizers than we.

In this section we'll take a brief foray into ways of making our speech recognizers more intelligent in detecting and correcting errors. Since recognizers are computers and making computer more intelligent belongs to the field of artificial intelligence, we will be examining one aspect of this burgeoning and promising field.

3.13.2.1 Syntax

We already examined how to use syntax to reduce perplexity, hence reducing the probability of error. (We called the process "subsetting.") Syntax is concerned with what words may co-occur with what words. When illegal co-occurrences are recognized, it is often due to speech recognition errors, since humans, for the most part, speak syntactically well-formed utterances.

Subsetting exploits syntax by reducing the active vocabulary that immediately follows a recognized word. There are two limitations to its usefulness. One is that if the current word is misrecognized, its influence on the next word under subsetting may increase the chance of error. Second is that many syntactic constraints operate over a distance rather than on adjacent words.

Let's look at our previous example using the words *the, old, man, men, walk, walks, sleep, sleeps.* Suppose the input to the recognizer is *the old man walks* but the recognizer substitutes *men* for *man.* With subsetting in place based on *men,* only *walk* and *sleep* are in the active vocabulary. *Walk* is the likely choice based on the input *walks* so the recognized sentence will have two errors in it: *the old men walk.* How could the machine detect these errors?

One way is to assign a confidence factor to both the word recognized and the runner-up, the word whose reference model is next closest to the input word. The same is done for the next word, disregarding subsetting. Here is what it may look like for just these two word positions:

    men (.75)   walks (.8)
    man (.25)   walk (.2)

This configuration is called a **word lattice.** Confidence factors for the four possible pairs are computed by multiplying the individual factors:

    men walks (.6)
    men walk (.15)
    man walks (.2)
    man walk (.05)

Men walks is the best guess with a score of .6, but the recognizer knows that it is syntactically ill-formed. This raises suspicion. Next, the system detects a dissonance: The original guess was *men walk* but *man walks* is more likely (.2 versus .15) when viewed in the larger context. The machine may alert the user to this fact or may override its first guess with its second guess in an action transparent to the user, who may blithely assume the recognizer got it all right in the first place. All this is part of **post-processing** in that it occurs after the initial recognitions. Such intelligent behavior is well within technological means.

Word lattices may be more extensive. They can be both deeper, meaning that first, second, and third best guesses are considered. Also, they can be wider, meaning that several words in a row may be considered. The principle, however, is the same as in our simple example.

Let's add the word *who* to our little vocabulary. See what a difference it makes! (And then imagining adding a few hundred words to catch up to the average four-year-old.) There are quite a few more grammatical sentences, for example *the man who walks sleeps, the men who sleep walk, old men who walk sleep,* and so on. The ungrammatical sentences are a little more interesting, too, such as *the men who walks sleep* and the like. The same syntactic constraints of noun subjects agreeing with verb predicates on which we based subsetting still pertain, but they no longer occur solely between adjacent words.

There is still a role for subsetting. All the previous subsetting constraints still hold. In addition, we can restrict the active vocabulary following *who* to *walk, sleep, walks, sleeps.* Even so there are still plenty of ungrammatical sentences not prevented by subsetting. If we assume that the input is well-formed, ill-formed output suggests errors in recognition.

To detect syntactically incorrect output during post-processing, the recognition requires access to a **grammar.** The grammar is a schema for designating all and only the syntactically well-formed sentences drawn from a vocabulary of words. There are many such schemata. One is a state transition diagram called a **transition network,** shown as Figure 3.17.

Only grammatical sentences are found by following the arcs (arrows) from state to state (the lettered rectangles) and noting the words on the arcs, if any, and proceeding to the final state K. If there are several words on an arc, any one may be chosen. If there are several arcs to the next state, any one may be chosen. If there are no words on an arc, the next state may be accessed without a word. Here are some sample sentences along with a state trace:

1. *The old man sleeps:* A B D F K
2. *The man walks:* A B D F K
3. *Men walk:* A C E K

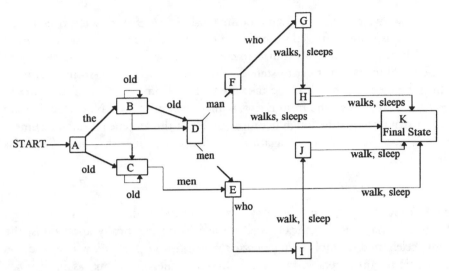

Figure 3.17 A transition network.

4. *The old, old, man who walks sleeps:* A B B D F G H K

5. *Old men sleep:* A C E K

In (1) we get *the* when we go from state A to state B; *old* from state B to state D; *man* from state D to state F; and *sleeps* is chosen in going from state F to state K. Had we chosen *walks,* the sentence would have been *the old man walks.* (2) is similar except we take the null arc (with no words) to get from state B to state D. Same for (3) where we take the null arc from A to C. In (4) we take the self-arc from B to B for the first *old,* and the arc from B to D for the second *old.*

The network can be viewed as a sentence generator. As you go from state to state you produce a word of the sentence. The string of words accumulated when you reach state K will be a grammatical sentence.

It can also be viewed as a sentence checker. You start with a putative sentence and try to move from state to state according to the words in the sentence. You can only leave a state on an arc corresponding to the word of the sentence being looked at currently, or on a null arc. If you succeed in reaching state K at the same time that you reach the end of the sentence, the sentence was syntactically correct; if you are stymied anywhere, the sentence was not syntactically correct. For example if we are examining *the old men who walk sleep* we would go from A to B and consume, or check off, *the.* From B we are allowed to go to D because *old* is on the B-to-D arc. The *old* is consumed. D to E consumes *men;* E to I consumes *who;* I to J for *walk* and on to K for *sleep.*

If we were checking the ungrammatical *the old men who walks sleep* we'd get as far as state I after consuming *the old men who.* However, we can't leave state I because the only way out is to consume *walk* or *sleep,* but the next word is *walks.* Since we can't get to state K, the input is deemed ill-formed. If we're checking *men sleep old* we could take the null arc from A to C, consume *men* in getting from C to E, and from E to K while consuming *sleep.* Now we're in state K but that's not good enough because the rest of the sentence is not consumed, in this case the word *old,* so again the system determines that *men sleep old* is ill-formed.

3.13.2.2 Semantics

Most communication has both syntax (structure) and semantics (meaning). Sentences may be syntactically well-formed but semantically anomalous, the most celebrated example being *colorless green ideas sleep furiously.* They may be semantically interpretable but not follow the rules of syntax, as in *the men walks.* It is generally easier to check sentences syntactically than semantically because many rules of syntax can be formalized, hence programmed. Semantics is a slippery subject. Sentences that on the surface appear semantically ill-formed often turn out to be perfectly interpretable in context, as in *I'm the soup, he's the salad;* the utterance appears to be semantically anomalous, but it made perfect sense to the cashier at the restaurant where my friend and I were paying our own share of the tab.

Imagine a speech recognition system that accepts dates in the form "month, day, year." To keep it simple, the vocabulary has the names of the twelve months, *January, February, . . . December,* and the digits *zero, one, . . . nine.* Lincoln's birthday would be *February one two one eight zero nine.* Independence day is *July zero four one seven seven six.* The syntax of these inputs is easy to describe. Its transition network is seen in Figure 3.18.

In Figure 3.18, the word *Month* stands for any one of *January–December,* and *digit* stands for any one of the digits *zero–nine.* This network accepts all the possible dates, including some that we don't want, such as *May three nine one*

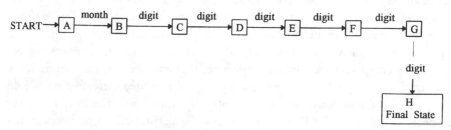

Figure 3.18 A transition network for dates.

nine nine eight (May 39, 1998). These, "bad" dates are syntactically well-formed, but semantically anomalous—they don't make sense.

Natural language speech recognition can be helped both by syntax and semantics. *Tom stole a big fat big* would be flagged as a syntax error, since a noun is required to end the sentence but *big* is an adjective. *Tom stole a big fat wig* is semantically anomalous as wigs are not ordinarily thought of as being fat. Both suggest recognition errors. Neither the syntax nor semantics of natural languages are understood deeply enough to allow their general use in unrestricted dictation systems, where the syntax may be correct but convoluted, and the semantics questionable but ultimately okay, as in *Tom stole a big fat wig to wear in the play.*

3.13.2.3 Context

Returning to our little example of inputting dates by speech, suppose we know that the dates are birthdays of living people. Then any date whose year is in the future, or too far back in the past, is likely to contain a recognition error. This constitutes the use of context. *June one five two nine six seven* (June 15, 2967) is syntactically and semantically well-formed—it's a legitimate date—but contextually ill-formed within the birthday context.

In the case of Tom stealing things that end in *-ig*, if the context is such that Tom is a hungry vagabond, *pig* or *fig* is to be expected; if context indicates that Tom quit smoking forty eight hours ago, maybe *cig;* if he's a James Bond type go for *MIG;* a hijacker, *rig;* a cross-dresser, *wig.*

Context provides the best speech recognition clues for humans and is the hardest to implement on a computer. The successful speech recognition systems of the world operate within a highly constrained context so that at least some of the contextual information can be encoded and used in error detection and correction.

3.13.2.4 Expectation

Expectation refers to the notion that we expect people to say certain things and not to say certain things, in a given situation. If we ask a question about the amount of the national debt, we expect a number, not a color in the answer. If someone asks you to describe what you wore to the president's inaugural ball he or she expects to hear about clothes, not spaceships.

Subsetting is sometimes based on expectation. After a third person singular subject we expect the verb to have an s on it. Those are the verbs in the active vocabulary. The concept of expectation, and its use in error detection and correction, takes place at all levels. Humans employ it all the time in going about their daily recognition of speech. Expectations are molded by meaning and context so this section overlaps with preceding ones.

If the person in the line in front of you at the grocery checkout turns and says to you "the Magna Charta was signed in 1215" you might well and rightfully say, "duh, what?" not because you're hard of hearing or zoned out on lurid tabloid headlines, but because the remark is so completely unexpected. On the other hand, if the person says, "The bagger is more slow today than she usually ever is," you'd probably be okay even though the latter sentence is arguably more difficult to recognize, what with all its monosyllabic words and unhelpful syntax.

When you call the phone company's speech recognizer, it expects you to talk about phone calls. *Collect call, long-distance call, reverse the charges, credit card call, operator, information,* and *directory assistance* are all utterances the system expects to hear. *Labrador retriever, a bunch of radishes, the Magna Charta* are among the many things it does not expect to hear. The recognizer is trained to spot words about placing phone calls, not words about, dog shows, grocery shopping, or English history. The epitome of expectation is when the recognizer advises you to "answer *yes* or *no*" and then prepares itself to spot one of those two syllables.

Smith and Hipp describe a dialog system in which expectation plays a major role [15]. The general context of the dialog is an electronic repair task. The user attempts to repair a broken circuit by conversing with the computer, which is programmed with knowledge of the circuit. As the task proceeds from diagnosis to repair, the computer expects different kinds of interaction. For example in performing the diagnosis the computer asks, "Is there a wire between connector 84 and connector 99?" The user answers, "It is not there," which is recognized as *in is not are*! Recognition errors are inferred based on the fractured syntax. Because *it* is easily misunderstood as *in,* and *there* is easily misunderstood as *are,* and the system expects a response regarding the existence of a wire, it is able to reconstruct the original utterance and respond appropriately to it.

Syntax, semantics, and contextual checking, together with expectation, can alert the recognizer to the possibility of an error. These are detection methodologies. The correction process may be to notify the user, to form a word lattice and choose the grammatical sentence of greatest confidence, or to employ phonetic similarity, confusion matrices, and statistical language models. The latter three topics are taken up in Sections 3.13.3.1-3.13.3.3.

3.13.3 Automatic Error Correction

When errors are detected most systems alert the user, who may exercise the various options for error correction, many of them described above. Ideally, we'd like the computer to provide both detection and correction, making accurate recognition as transparent to the user as possible.

One method of automatic correction is to compare each word in the flawed output with all other vocabulary words in an attempt to undo possible substitution errors. Likely candidates can be substituted back into the utterance, which is then reexamined for syntactic, semantic, and contextual consistency, and if feasible, given a score for comparison with other candidates.

3.13.3.1 Phonetic Similarity

Substitution errors occur most often because of phonetic similarity. A table can be created in which each vocabulary word is at the head of a column of phonetically similar words ranked as to degree of similarity. For our toy vocabulary used above (see Figure 3.17), such a table might look like Table 3.8.

If *the man old walks sleep* is recognized and errors detected, the system could try to correct them by making trial substitutions from the table, beginning with the first row under the vocabulary words because they contain the most phonetically similar elements. It might try substituting *sleep* for *the* as a start. A syntax check rejects that possibility. Substituting *men* for *man* doesn't help at this point, but the substitution of *who* for *old* improves the situation. Substituting for *walks* is no help, but *sleeps* for *sleep* gives as the grammatical sentence *the man who walks sleeps.* We can't quit here, however. All possibilities need to be considered and there is one more, namely *the men who walk sleep.* To rank these two choices we would need to know whether *sleeps* is more phonetically similar to *sleep* than *men* is to *man.* Thus it would help to know the **degree** of similarity, information lacking in Table 3.8.

The matter of determining the degree of phonetic similarity is complex, though it can be done through phonetic analysis. We won't attempt to complete the job, but the idea is this. *Pig* is more phonetically similar to *big* than to *dig* because both /p/ and /b/ are bilabial stops, but /d/ is an alveolar stop. Further, *pig* is more similar to *dig* than to *cig* because both /p/ and /d/ are stops, but /s/ is a fricative. One more: /ă/ is more similar to /e/ than to /u/ because both /ă/ and /e/ are relatively low front vowels while /u/ is a relatively high back vowel. We might also argue, a little tenuously I confess, that /ă/ is more similar to /e/

Table 3.8
Phonetically Similar Words

Man	*Men*	*Old*	*Sleep*	*Sleeps*	*The*	*Walk*	*Walks*	*Who*
Men	Man	Who	Sleeps	Sleep	Sleep	Walks	Walk	Old
		Walk	The	The	Sleeps	Old	Old	
		Walks						

than "null" is similar to /s/, and therefore *man* is more similar to *men* than *walk* is to *walks*. Based on that analysis, we could decide that *the men who walk sleep* is the best choice to correct the misrecognized *the man old walks sleep*.

3.13.3.2 Confusion Matrices

A **confusion matrix** tabulates the performance of a speech recognizer. The statistics may be accumulated for a single session from a single speaker, or from multiple sessions with one or more speakers. It is a table of numbers whose row and column headings are the vocabulary words of the speech recognizer. The number in row i, column j, is the number of times word i was spoken, and word j was recognized.

A confusion matrix gives a history of substitution errors based on experience. The errors may be related to phonetic similarity but often they are not. They may be due to the faulty creation of reference models or templates, or to the speech of a particular user, or to an idiosyncrasy of the recognizer itself. Figure 3.19 is a confusion matrix that was created in my laboratory during a study of neurologically disturbed speech.

The experiment was based on the discrete word recognition of a 20-word vocabulary, given as row and column heads in Figure 3.19. The word labeling the row is the word spoken. Each word was spoken 45 times. The number in the column is the number of times the word was recognized as the word at the head of the column. For example *zero* was recognized as *zero* 35 times, as *flip* four times, as *silk* three times, as *yourself* one time and as *enjoy* one time. One time it was rejected, as indicated in column 21. A blank cell means zero, so *smooth* was never substituted for *zero*.

The analysis summarized in the confusion matrix is interesting from several points of view. The numbers along the diagonal are a measure of how well a word was recognized, the larger the better. *Smooth* did the worst, being recognized correctly only 25 out of 45 times. If this were a real application, we might try to find a substitute for *smooth* in the vocabulary. Four words were recognized perfectly: *although, understand, enjoy, has*.

Let's compare monosyllables with polysyllables. The monosyllables are *smooth, weigh, flip, through, pushed, child, silk, blink, has*. They are recognized correctly 79% of the time. The polysyllabic words are recognized correctly 95% of the time.

It is also apparent that the substitution errors are much more varied for monosyllabic words than the others. Eight different words are substituted for *flip*. On the other hand, the only substitution for *conversation is understand*.

The confusion matrix can be used for error correction in much the same way as the table of phonetic similarity. If errors are detected in an output, substitute back the most commonly substituted word. For example, if *understand*

Word Presented \ Word Matched	zero	smooth	weigh	about	flip	through	beautiful	pushed	child	programmer	silk	measure	yourself	conversation	blink	glowing	although	understand	enjoy	has	Not recognized
zero	35				4						3		1						1		1
smooth	1	25				7	4				2	4			2						
weigh			38	1			1								1	1					3
about				39	1						1										4
flip	2	1			26		2	1	2				1			1	7				2
through		10			1	30									1	1					2
beautiful		1					43				1										
pushed			1		1			37							5						1
child				1					42							1					1
programmer				1						41						3					
silk	1			1				1			39	1							1		1
measure								1				44									
yourself				1									42								2
conversation														39				6			
blink	1	1						1							39	2					1
glowing		1														44					
although																	45				
understand																		45			
enjoy																			45		
has																				45	

Figure 3.19 A confusion matrix for speech recognition.

was in the recognized sentence and there was a problem, we might replace it with *conversation* to see if matters improved.

For systems with multiple users, a confusion matrix can be created for each user. This would be part of the user's profile, or **user's model**, and exploited to optimize error correction on a personal basis.

3.13.3.3 Statistical Language Models

A confusion matrix based on a history of usage is one kind of statistical model. **Statistical language models** are built by observing word occurrences in millions of sentences collected from a wide variety of written and spoken sources. Their principal use is in large vocabulary systems that take general dictation. They are also called, simply, **language models.**

Statistical language models are based on the following fact: The likelihood that a certain word will occur in a sentence is related to the words that precede it. For example the likelihood that *fleas* will occur in a sentence is higher when the previous words are *my dog has* than when the previous words are *my boss eats.*

Statistical language models come in various sizes. Trigram language models determine the probability of a third word, given the first two words. These probabilities are measured empirically by examining occurrences of those first two words and seeing what the third word is in each case.

Let's do it. Consider the last three paragraphs. The words *language models* occur five times (not counting the section heading). The two words are immediately followed by *are,* ⟨*period*⟩, *are, come, determine.* These words, especially *are,* would be given preference in a speech recognition event in which *language models* was recognized and there was some question about the word that followed.

An *N*-gram model would be based on determining the probabilities of an *N*th word, given the first $N - 1$ words. Thus 4-gram, 5-gram, and 6-gram models are possible, but there is evidence that little useful information is gained beyond the trigram size [16]. It is also true that a trigram model is more effective than a bigram model, which looks only at the preceding word. However, combining both bigram and trigram models on a three-word sequence is better than the trigram model used alone. If the three words are W1, W2, and W3, combining the probabilities of W2 given W1; W3 given W2, and W3 given the two preceding words W1 and W2, leads to better results.

The models can be used both for reducing errors by reducing perplexity, and for error correction after an error has been detected but not pinpointed. If the recognizer thought it heard *language models there useful,* the phonetic similarity of *there* and *are,* together with the trigram model, would suggest that the sentence actually spoken was *language models are useful.*

Language models can be specialized to particular users and particular applications. Individuals have their own style of writing in which certain word patterns occur more frequently than others. These comprise a personal language model which may be incorporated into the overall user model for that individual.

Clearly, word patterns vary according to what is being written. A computer usage manual will differ from a physician's report or a mystery novel. The language model can also incorporate this fact.

One major drawback to statistical language models is the truly vast amount of data that must be collected to establish the model's statistical significance. A vocabulary of *N* words permit N^3 distinct trigrams. For a vocabulary of 20,000 words, a bare minimum for general dictation, there are eight trillion possible trigrams. Collecting and processing a body of data large enough to be of practical use is daunting. In practice, one uses relatively "small" amounts of data (250,000,000 words!) and draws the best conclusions possible. (See [16] for a more complete discussion.)

Another drawback to statistical language models is that there are numerous ungrammatical sentences—*you be heard as better results are poor and lonely in the capacity*—in which any three words in a row may be found in grammatical sentences. Compare it is important that you be heard; she will be heard as long as she is able to speak; the music was heard as better on this radio; they used this protocol as better results were obtained; etc.

Despite the problems, language models are used today, in various forms, by many successful automatic dictation systems such as IBM's *ViaVoice®* and Dragon Systems' *Naturally Speaking®*.

3.14 Future Challenges II

The accurate recognition of naturally spoken speech in a general context is an unachieved goal and remains the primary aim of speech recognition research. This recognition should be of speech spoken in a typical daily environment such as a busy office and should not require speakers to wear a microphone— what most of us do on a daily basis without thinking much about it.

When this challenge is met, other even more daunting challenges will appear. The automatic translation of one continuously spoken language into another in approximate real time—what a United Nations interpreter does—will appear high on the list. The seeds of this application are already sprouting. Long-distance companies provide the automatic translation of single words from one language to another in order to help automate international telephone calls.

Another challenge will be to recognize the speech of one individual speaking among several other talkers, the way you focus on recognizing the speech of that attractive person you're speaking with at a noisy party.

Ultimately, the challenge is to recognize all the speech of all the people speaking simultaneously. At that point, our computers will have exceeded our own abilities.

Though you may not be a speech processing professional, you will be able to gauge progress in speech recognition throughout your lifetime by observations in three domains, the first of which is communications: Communications companies have invested heavily in speech recognition. They see the technology both from the point of view of saving labor costs and expanding communications options. You can monitor progress in speech recognition by keeping up with the voice options offered by your telephone companies.

The second domain is software for your personal computer. At the end of the millennium we find speech recognition becoming a standard option on personal computers. The quality of the speech recognition that is offered will follow the state of the art very closely.

Most of you have seen a court stenographer at work, either in real life, or on one of the many TV shows or motion pictures that depict courtroom scenes. When a speech recognizer takes over this task, if you're alive, you'll know that speech recognition has truly arrived.

References

[1] Rodman, R. D., "Computer Speech Recognition," in *Language, Speech and Mind*, pp. 269–294, L Hyman and C. Li (eds.), New York, NY: Routledge, 1988.

[2] Crevier, D., *AI: The Tumultuous History of the Search for Artificial Intelligence*, New York, NY: Basic Books, 1993.

[3] Turing, A. M., "Computing Machinery and Intelligence," *Mind*, Volume 59, 1950, pp. 433–460.

[4] Lee, K. F., *Automatic Speech Recognition*, Norwood, MA: Kluwer Academic Publishers, 1989.

[5] Schafer, R. W., and J. D. Markel, *Speech Analysis*, New York, NY: IEEE Press, 1979.

[6] Klatt, D., "Review of the ARPA speech Understanding Project." *Journal of the Acoustical Society of America*, Vol. 62, 1977, pp. 1345–1366.

[7] Winograd, T., *Language as a Cognitive Process*, Reading, MA: Addison-Wesley Publishing Company, 1983.

[8] Biermann, A., R. Rodman, and F. Heidlage, "Natural Language With Discrete Speech as a Mode for Human to Machine Communication," *Communications of the Association for Computing Machinery*, V.28, No. 5, 1985, pp. 628–636.

[9] Rabiner, L. and B. H. Juang, *Fundamentals of Speech Recognition*. Englewood Cliffs, NJ: Prentice-Hall, Inc., 1993.

[10] Dahl, H., *Word Frequencies of Spoken American English*, Essex, CT: Verbatim, 1979.

[11] Rabiner, L., "A Tutorial on Hidden Markov Models and Selected Applications in Speech Recognition," *Proceedings of the IEEE*, V. 77, No. 2., 1989, pp. 257–286.

[12] Markowitz, J. A., *Using Speech Recognition*, Upper Saddle River, NJ: Prentice Hall PTR, 1996.

[13] Tetschner, W., *Voice Processing, Second Edition*, Norwood, MA: Artech House, Inc., 1993.

[14] Rodman, R. D., M. Joost, and T. Moody, "Performance Evaluation of Connected Speech Recognition Systems," *Proceedings of the Speech Tech '87 Voice I/O Conference*, New York, NY: Media Dimensions, Inc., 1987, pp. 269–274.

[15] Smith, R. W., and D. R. Hipp, *Spoken Natural Language Dialog Systems*, New York, NY: Oxford University Press, 1994.

[16] Jelinek, F., "The Development of an Experimental Discrete Dictation Recognizer," *Proceedings of the IEEE*, Vol. 73, No. 11, 1985, pp. 1616–1624.

4

Speech Synthesis

Speak clearly, if you speak at all; carve every word before you let it fall.

Oliver Wendell Holmes, Sr.

4.1 Introduction and History

Speech synthesis is concerned with making computers talk. Since many modern devices, such as the automobile, contain computer chips, speech synthesis is also about giving them voice.

There are two aspects to speech synthesis: the actual physical process of producing speech sounds—of making a machine vocalize—and, second, telling it what to say—in effect, teaching it to read some set of symbols.

What's older, the first talking machine, or the United States of America? The answer is the United States, but not by much. While our forefathers were bringing forth on this continent a new nation, Christian Gottlieb Kratzenstein was bringing forth a new machine, "an instrument constructed like the vox humana pipes of an organ which . . . accurately express the sounds of the vowel."[1] The year was 1779. Kratzenstein's machine consisted tube-like acoustic resonators of various shapes, each of which made a vowel sound when set to resonating by a reed forced to vibrate in a stream of air.

A few years later Wolfgang von Kempelen constructed a more elaborate talking machine in Vienna. This device used bellows to produce an airstream and constricted passages controlled by the fingers to simulate consonantal

---

1. Quoted from the question posed by the Imperial Academy of St. Petersburg as subject of their annual prize, which Krazenstein won.

sounds. A conical resonator of leather, whose shape could be adjusted by hand, produced vowel sounds. Von Kempelen demonstrated his machine in 1791.

A replica of von Kempelen's machine inspired a young Alexander Graham Bell to build the talking head mentioned in Chapter 1. Together with his brother Melville, the boys made a cast from a human skull. Into it they inserted vocal parts made from wire, rubber, cotton and wood to replicate the lips, teeth, tongue, hard palate, velum, and pharynx. The vocal cords and larynx were simulated by rubber bands in a tin box, and of course bellows served for lungs. They controlled this early speech synthesizer with a system of levers. It was capable of producing short utterances by combining various speech sounds.

The mechanical approach to speech synthesis was aptly summarized in *Scientific American* in 1871.

> Machines which, with more or less success, imitate human speech, are the most difficult to construct, so many are the agencies engaged in uttering even a single word—so many are the inflections and variations of tone and articulation, that the mechanician finds his ingenuity taxed to the utmost to imitate them.

Though others tried to improve on the Bells' model, progress would have to wait for the electronic age.

In the 1920s an engineer built a formant synthesizer consisting of an oscillator and two electrical resonators. It was able to reproduce vowels, nasals, and a few words such as *mama*. You may recall the **vocoder**, discussed in Chapter 2. It recorded speech in an abridged format that could later be played back. With a few modifications it could synthesize speech electronically. An operator manipulated keys to simulate the acoustic parameters of the speech that was to be synthesized. The **voder**, as the vocoder was then called, was a big hit at the 1939 New York World's Fair. It was barely intelligible, but its real drawback was that the operator required about the same amount of training as a concert pianist.

All of the synthesizers discussed so far required human operators. In that sense, they were more like musical instruments than speech automatons. Only in the 1950s were scientists able to synthesize the vocal tract electronically and produce hands-off devices that accepted symbolic, electronic input and returned intelligible voice as output. The invention of, and the improvements to, the integrated circuit led to successively higher quality electronic speech between the 1960s and the present. Figure 4.1 is a model of today's speech synthesis process, centered on a device called a **speech synthesizer.**

These same circuits gave rise to computers. Computers permitted a new kind of "speech synthesis" not previously possible: Individual words could be

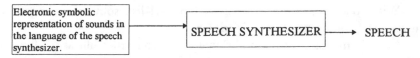

Figure 4.1 Model of a speech synthesis process.

recorded from a human voice and then rearranged by the computer to form varying outputs. Individual recordings of the digits, *zero, one, . . . nine,* plus the phrase *area code,* could be rearranged to speak all possible phone numbers in North America.

In the remainder of this chapter we will distinguish between **parametric coding,** the electronic synthesis of speech "from scratch"; and **concatenative synthesis,** the rearrangement of prerecorded human speech.

4.2 Parametric Coding (Electronic Synthesis)

Parametric coding, also called **rule-based synthesis,** is speech synthesis by recipe. The cooking metaphor is apropos. To prepare a dish in the kitchen, you assemble the ingredients (parameters), modify them with knives, whisks, sifters, cold, heat, etc., and combine them in specific proportions. To prepare a synthetic vowel sound, one begins with an electronic tone at the frequency of vibrating vocal cords, and all its harmonics, as ingredients. Filters modify the harmonics to accentuate the formant frequencies, and de-emphasize other frequencies, in the proportions specific to the vowel being synthesized.

Recipes for consonants require additional ingredients and modifications. Sound must be cutoff for stops and affricates, resonances added for nasals, hissing and buzzing noise tossed in for sibilants like /s/ and /z/, and white noise (aperiodic sound) blended into the mixture in just the right proportion for voiceless consonants.

Cooking is time-dependent. "Roast 20 minutes per pound," "stir for five minutes," or "baste every half hour." Speech synthesis is highly time-dependent, and the proportion of ingredients must be adjusted frequently. Indeed, the dynamic character of speech sounds requires that parameters be updated hundreds of times per second, typically, once every five milliseconds.

4.2.1 Parameters of Parametric Speech Synthesis

A well-stocked kitchen contains the basic ingredients of cooking such as flour, sugar, salt, and spices. Likewise, a speech synthesizer contains circuits that simulate the parameters of human speech sounds, the building blocks out of which all sounds may be constructed.

With no attempt to be exhaustive, Table 4.1 lists some of the ingredients that go into cooking up electronic speech.

Many synthesizers have additional parameters that allow small adjustments to the voice quality, and these may differ from synthesizer to synthesizer, just as the contents of the spice cabinet will differ among chefs. The parameters given in Table 4.1 are essential, more or less the parsley, sage, rosemary and thyme of parametric speech synthesis.

Every synthesizer has a vocabulary of speech sounds—its menu—that it is capable of speaking. It takes as input a symbol representing a unit of sound. It looks up the recipe for that sound and assembles the parameters to produce an electronic representation of the sound acceptable as input to a speaker or telephone line.

Synthesizers, then, are language-specific. They accept the speech sounds of the language for which they are designed. An English synthesizer would be capable of producing the allophones of English that we saw in Chapter 1. A synthesizer for a language other than English would be essentially the same from an electronic point of view but would accept a different set of input symbols appropriate to its own phonetic inventory. French would have nasal vowels (*bon*); German would have a velar fricative (*Bach*); Russian a palatal nasal (*nyet*), Italian an alveolar affricate (*pizza*), etc. Moreover, synthesizers for tone languages would need the an input vocabulary that would allow the correct tone to be placed on the vowel.

Most synthesizers accept additional input secondary to the speech. The sounds remain the same, but the final product varies along different lines. For example the rate of speech may be varied from slow to fast, corresponding to speaking rates between 100 and 400 words per minute. Synthesizers will accept input characters that tell it to remain silent, as in the pause that precedes the release of a stop consonant. They will accept symbols that determine changes in pitch, so that the voice isn't a droning monotone, and they will accept durational symbols that dictate how long a particular sound should be prolonged (or foreshortened). As a result, many synthesizers can, and do, sing, and some can articulate tongue twisters like *the sixth sheik's sixth sick sheep* faster than any human.

Some synthesizers accept metasymbols that result in adjustments to a panoply of parameters. For example DECtalk®, a brand of speech synthesizer manufactured by Digital Equipment Corporation, is capable of speaking in the guise of a man, woman or child. Dozens of parameters would be slightly altered to convey the impression of gender and age. The first generation of DECtalk® boasted six voices, named Rough Rita, Frail Frank, Huge Harry, Perfect Paul, Beautiful Betty, and Kit the Kid. Rough Rita was intended to be an older, heavier woman's voice, while Beautiful Betty's was supposedly the voice of a

Table 4.1
Some Parameters of Electronic Speech Synthesis

Parameter	Comments
Fundamental frequency of voicing	Simulates the vibrating vocal cords. Ranges from 100 Hz (a man with a deep voice) to 450 Hz for the voice of a female or child
Amplitude of voicing	Determines how loud the synthesizer speaks
Nasal frequencies	These parameters replicate the effect of nasality; they are needed not only for [m, n, ng], but to tinge vowel sounds with nasality when they precede a nasal sound; they may also be adjusted to give or deny a "nasal twang" to the synthesizer
Nasal amplitude	Determines amount of nasality perceived
Frequency of nth formant	Center of frequency band of first, second, etc. formants (see Section 1.3 for ranges); all synthesizers have values for $n = 1, 2, 3$; many include the 4th and 5th formants as well
Bandwidth (frequency ranges) for nth formant	Determines the bandwidth of the formant; for example, in Figure 1.5, the spectrogram for [ē]; the frequency range of the first formant is between 100 and 450 Hz
Amplitude of nth formant	The decibel output of the various formants controls subtle aspects of speech quality
Frequency range of fricatives	Although these sounds are aperiodic (white noise), they occupy a frequency range between 2,000 and 15,000 Hz
Amplitude of fricatives	Must be balanced with other speech components
Amplitude of aspiration	Aspiration is, in effect, a glottalic fricative (much like [h]) and is in the frequency range of fricatives but with a different amplitude
Degree of breathiness	Breathiness is a measure of the amount of air that leaks past the vocal cords without contributing to their vibration; it is important for imparting distinguishing characteristics to the voice; for example, female voices have, in general, more breathiness than male voices
Degree of laryngealization	This is the amount of creakiness in the voice; it results from very low frequency vocal cord vibrations and generally occurs at the ends of words; like breathiness, this feature imparts characteristics to the voice that are related to its perceived sex and age

younger, slimmer woman. (This was the early 1980s, when "PC" still stood for personal computer not political correctness.) Frank was an older man of slight build with lots of breathiness in his voice. Harry was a heavy, large-headed man with the deep voice of a basso profundo. Paul was the default voice, a man of average height, weight, and age. Kit was intended to be a prepubescent male voice. One could also adjust parameters to create a custom voice.

DECtalk® would also accept metasymbols that adjusted suprasegmental features such as pitch, loudness and duration to achieve **prosodic** effects, both on word and phrase levels. (**Prosody** refers to the combined effect of suprasegmental features, as discussed in Section 1.2.4. The synthesizer could distinguish between *inSERT* the verb and *INsert* the noun by varying the stress within the word. It could even accommodate the subtle intonational differences that distinguish the ambiguous *he left instruction for her to follow* also discussed in Section 1.2.4.

4.2.2   Input Units of Parametric Speech Synthesis

With the ingredients listed in Table 4.1, we could in theory replicate any speech units, from allophones through sentences. If a particular synthesizer was limited to a finite number of English sentences such as *At the tone the time will be twelve o'clock exactly, At the tone the time will be twelve o'clock and ten seconds, At the tone the time will be twelve o'clock and twenty seconds, . . . At the tone the time will be eleven fifty-nine and fifty seconds,* they could be prepared in advance and numbered 1–4,320. The synthesizer's input would be the index number, and its output the synthetically spoken corresponding sentence. The idea is silly for at least two reasons. First, if the needed sentences were known in advance, their words could be prerecorded by a person more easily and economically and then given to a concatenative synthesizer. Second, human languages are capable of expressing infinitely many sentences. A sentence-based synthesizer could never realize this potential.

An improved choice of units might be words, but that is scarcely more practical than sentences. Taken in all possible grammatical forms—*eat, eats, ate, eaten, eating*—there would be millions of word recipes to prepare, and the system would have to be continually updated as new words are coined such as *e-mail,* and parents invent new names for their children.

For reasons of sheer size, morphemes (hundreds of thousands), syllables (10,000), demisyllables (thousands), and diphones (thousands) are not feasible units. (Morphemes, demisyllables, and diphones are discussed in Sections 4.3.2 and 4.3.3.) That leaves allophones (dozens to hundreds). Most parametric speech synthesizers accept allophones as input. Usually, they are buffered and processed in groups, so that a few general purpose smoothing adjustments can be applied at the boundaries to minimize inter-allophonic effects. This obviates to some degree the need for diphones.

The phonetic alphabets of Chapter 1, given as Tables 1.3 and 1.4, approximate the input alphabets of most parametric speech synthesizers quite well. For example, our alphabets have a combined total of 45 units. An early version of DECtalk® also had 45 allophonic units, though not exactly the same ones. In addition DECtalk® accepted about a dozen symbols that controlled various elements of prosody, such as the single quotation mark, which was interpreted as primary stress.

Parametric synthesizers have never sounded natural, though they have reached high levels of intelligibility. One reason for the lack of naturalness is that they make use of too few allophones. Arguably, English has infinitely many allophones. Each vowel could be articulated slightly differently along a continuum of tongue position and lip rounding. Similarly, voiceless stop consonants could vary ever so slightly along a continuum of degree of aspiration. It's not clear how many allophones are needed to achieve natural-sounding English. Certainly not infinitely many, and probably not even a thousand, but definitely more than the few dozen that most parametric synthesizers provide.

At a minimum, the voiceless stops and affricates require three allophones each: one aspirated, one unaspirated, and one unreleased for the ends of syllables. The voiced stops and affricates require two allophones, released and unreleased. Vowels require more allophones than you might think at first blush, even disregarding small changes in tongue position. Every vowel needs an oral allophone, a nasal allophone for co-occurrence with nasal consonants, and a durationally longer allophone for syllables whose final consonant is voiced. (The vowels in *beet* and *bead* differ measurably in length.)

There are numerous other small allophonic variations that determine whether speech sounds natural and human-like, or machine-like. I'll mention one more. The velar stop /k/ has a perceptibly different allophone when preceding an /ē/ than when preceding a /ū/. The /k/ in *keel* and the /k/ in *cool* are not pronounced identically. The /k/ is palatalized before /ē/, that is, the stop closure occurs closer to the hard palate than otherwise. It is similar for /g/. Using the same allophone in all phonetic environments sounds unnatural. To fully accommodate the palatalization, three more allophones are needed for /k/ and /g/. Table 4.2 summarizes the allophones for the velar stops.

The second reason that parametric synthesizers—and other types of synthesizers to be discussed in Section 4.3—sound unnatural is the prosody is often unnatural. Neither stress, nor segment length, nor syllable length, nor pitch, nor intonation are quite on the money all the time. This is not a problem with the circuits that control these parameters; they are capable of any degree of nuance. It is a problem of determining precisely the right settings at the

Table 4.2
Some Allophones of /k/ and /g/

Allophone	Context
k^h (aspirated)	Initial in stressed syllables before non-front vowels (e.g., *kudos, could, because*)*
k (unaspirated)	After /s/ before non-front vowels; syllable initial in unstressed syllables before non-front vowels; syllable final position, sometimes** (e.g., *skull, scoot, teacup, peak*)
k$^=$ (unreleased)	Syllable final position, sometimes (e.g., *attic*) (see previous footnote)
k^{yh} (palatalized, aspirated)	Initial in stressed syllables before front vowels (e.g., *keep, caper*)
k^y (palatalized, unaspirated)	After /s/ before front vowels; syllable initial in unstressed syllables before front vowels (e.g., *skeet, monkey*)
g	Syllable initial before non-front vowels; sometimes syllable final position (e.g., *gulag, sheepdog, ago, argon*)
g$^=$	Syllable final position, sometimes (e.g., *rig*)
g^y (palatalized)	Before front vowels (e.g., *geese, regain*)

*These are not the only contexts. The matter is complex. For example the /k/ in *krypton* is aspirated even though it is followed by /r/.

**Speakers may or may not release a consonant in syllable final position. Both the aspirated and unreleased allophone occur in this context idiosyncratically. However, syllable final position is the only context in which the unreleased allophone occurs.

right time. The human ear is incredibly sensitive to any deviation from what it considers natural prosody, but the principles that govern natural prosody have proven as elusive as the perfect soufflé.

4.3 Concatenative Synthesis

It is a sign of the times that parametric synthesizers are becoming passé. Neither the allophone nor the prosody problem has yet been solved satisfactorily. Progress is being made but a new kid has blown into town by the name of **concatenative synthesis.**

Actually, the kid isn't new, but he has recently acquired a lightening processor, acres of memory, untold gigabytes of storage, and a dime store's price tag. Today it is feasible for the units of synthesis to be digitized, human speech. The job of the "synthesizer" is arranging these units into the desired output, adjusting prosody, and smoothing the boundaries between units to prevent infelicities in articulation, all of which are still worthy challenges.

Concatenative synthesis has been around since the 1950s, though in an uncanny passage in *Gulliver's Travels,* published in 1726, the satirical writer Jonathan Swift described the essential character of concatenative synthesizers:

> These bits of wood were covered on every square with paper pasted on them, and on these papers were written all the words of their language, in their several moods, tenses, and declensions, but without any order. The professor then desired me to observe, for he was going to set his engine at work. The pupils at his command took each of them hold of an iron handle, whereof there were forty fixed round edges of the frame, and giving them a sudden turn, the whole disposition of the words was entirely changed.

Originally, concatenative synthesis consisted merely of digitally recorded whole utterances, any one of which could be played back on demand according to the situational needs. If the spoken output consisted of a relatively small vocabulary of words, each one could be recorded digitally and played back in the proper order on request. A good example is *"At the tone the time will be . . ."* mentioned above. This method was especially effective for outputs such as numbers (phone, serial, account, etc.), dollar amounts, times of day, and dates.

The advantage is clear. The unit itself—syllable, word, etc.—has the clarity and naturalness of human speech within the limits of digitization (discussed in Chapter 2). In particular, the pronunciation of consonants, and especially consonant clusters such as *str-* or *-rsts,* which parametric synthesizers are poor at, is now as good as humanly possible. It is similar for the vowel durations in stressed syllables, which parametric synthesizers still can't get consistently right. Concatenative synthesis at the word level solved the allophone problem, and that part of the prosody problem concerned with the relative durations of segments within the word.

Two bugaboos of concatenative speech synthesis remain: interunit effects and prosody over and above the level of the recorded units. For example, in naturally spoken speech, words are run together. There are no "white spaces" except where the speaker pauses for breath or effect. Natural between-word articulation still needs to be simulated. Likewise, appropriate prosody has to be computed because it depends heavily on the context of the entire utterance, which isn't known when the speech units are pre-recorded. With little computing power, early concatenative synthesizers could only speak in a monotone with discontinuities between each word. You've likely heard concatenative synthesis from the phone company's directory assistance.

Until recently, parametric synthesis ruled the roost because it could handle a vocabulary of unlimited size. Modern computers now allow hundreds of thousands of words to be prerecorded digitally. The synthesizer rearranges them as needed, smoothes between words, and attempts to approximate the correct prosody, all in real time.

Nonetheless, for truly *general* output—the kind parametric synthesizers are capable of producing because their units are sound-based—one cannot use words as units. Apart from the impracticality discussed above, word-based synthesis is unable to handle the many applications that require a large unlimited vocabulary, such as ones that "read" texts not known in advance. (This will be discussed in Section 4.4).

4.3.1 Allophone Concatenation

Hypothetically, allophones appear to be the perfect unit for concatenation. A few hundred of them will serve as the basic building blocks for all utterances. The problems only appear upon implementation. How does one pre-record [k], say? It's impossible to articulate this consonant without also articulating a vowel, however short, as either the **onset** (the beginning part) or **offset** (the ending part).

Not to be stymied, speech engineers suggested that the allophones be isolated from longer utterances. From a recording of *Nick said he would skin one of his kin tomorrow,* they would attempt to isolate the three allophones of /k/, and use them in the synthesis of all words containing those allophones, such as *skim, skip, kite, come, tick, sick* and so on. This turns out to be both a good idea and bad idea. It's a bad idea because a major difficulty with allophone synthesis is articulation effects between allophones; concatenating individually produced allophones sounds unnatural because allophone offsets tend not to match up smoothly with allophone onsets. It's a good idea because isolating units from free-flowing speech lends a naturalness to the units. The trick is to select the right unit.

4.3.2 Diphone Concatenation

The *raison d'être* for diphones, whether used as units of speech recognition (see Section 3.6.4) or speech synthesis, is that they capture the articulation effects between allophones. Moreover, the articulation effects between the diphones themselves are not as serious a problem because the diphone boundaries are the midpoints of the same sound, which tend to match up fairly well. For example, to synthesize the word *scrimp* in *we scrimp and save,* we concatenate the diphones [ē-s], [s-k], [k-r], [r-i], [i-m], [m-p], and [p-ă]. (The diphone [ē-s] is the last half of [ē] plus the first half of [s]. It is concatenated with the diphone [s-k] consisting of the last half of [s] plus the first half of [k], which is concatenated with the diphone [k-r], and so on.)

Taking allophonic variation into account gives us the new concept of **allodiphone**—diphones made up of the combinations of the various allophones of the language. An illustration is the two allodiphones [i-kʰ] and [i-k], as in *liqueur* and *liquor,* containing two of the allophones of /k/, [kʰ], and [k]. The fact that there are thousands of such units is no longer a stumbling block due to advances in computer technology.

To achieve the most natural sounding allodiphones, they must be isolated from naturally spoken speech. Unfortunately one must process enormous amounts of speech to build a database of allodiphones sufficient for high quality concatenative speech. Today's diphone synthesizers compromise and use a modest number of allodiphones derived from artificially concocted free flowing speech.

4.3.3 Demisyllable Concatenation

Demisyllables occur as two types: the syllable onset plus the first half of the nucleus vowel sound and the second half of the nucleus vowel sound plus the syllable coda. (Refer to Section 1.2.5 for definitions.) They are illustrated in Table 4.3. ($V|$ denotes the first half of the vowel V; $|V$ denotes the second half of the vowel V; C denotes a consonant)

There are several thousand demisyllables, ignoring all but the most common allophones. If we get into allodemisyllables—that is demisyllables that are

Table 4.3
Possible Forms of Demisyllables

Form of Demisyllable	Example
$V|$ or $|V$	i
$CV|$	ki
$CCV|$	ski
$CCCV|$	skri
$|VC$	ik
$|VCC$	ilk
$|VCCC$	ilks
$|VCCCC$	irsts

composed of differing allophones (pi| and pʰi|, for instance)—that number becomes much larger.

Demisyllables have two strengths as units of concatenative speech synthesis. One is in consonant clusters such as the *skr* in *scrap*. While diphones are better than simple allophones in such clusters, demisyllables are best of all because the entire cluster is taken from human speech.

The other advantage of the demisyllable is in achieving natural sounding segment durations. Consider the two words spelled t-o-r-m-e-n-t. The noun has primary stress on the first syllable (*TORment*); the verb has primary stress on the second syllable (*torMENT*). The first syllable is longer in *TORment* than it is in *torMENT,* and the second syllable is longer in *torMENT* than in *TORment.* A synthesizer that fails to capture these nuances will be perceived as unnatural. Since the length of a syllable ending in one or more consonants is distributed over the vowel and the final consonant(s)—trust me, it's true—synthesizing natural sounding syllable lengths is more easily achieved with demisyllables than with other units.

A drawback of demisyllables is that not all syllable boundaries fit smoothly together. Consider a word such as *upkeep.* The *p*-to-*k* syllable boundary is a misfit that will be heard audibly and sound mechanical. With diphones the fit would be [p-k] concatenated with [k-p], a smoother transition.

Recently some speech engineers have proposed a mixed inventory of both diphones and demisyllables, taking advantage of their respective strengths and compensating for their respective weakness. Table 4.4 summarizes the principle advantages and disadvantages of the various units of concatenative synthesis.

4.3.4   Waveform of Concatenative Units

The units of concatenative synthesis are encoded digitized human speech. All of the methods of digitization discussed in Chapter 2 are potential candidates for use in synthesis. Two factors dominate the choice. One is the complexity of decoding—synthesizers must operate in real time. The second is the ease in which prosodic adjustments can be imposed on the speech units.

The decoding of LPC speech is the least complex, as we have seen. (The encoding process is the most complex and time-consuming, but it only needs to be done once.) However, that's not its chief advantage. The pitch of LPC speech can be adjusted via a single parameter, and the all-important duration of a speech unit can be adjusted by the simple expedient of adding or subtracting frames during decoding. Thus some, but not all, prosodic adjustments can be effected at the decoding stage.

Waveform coders require two stages for synthesis. One is the decoding process that produces the reconstructed waveform. The second is the imposi-

Table 4.4
Advantages and Disadvantages of Units of Concatenative Synthesis

Concatenation Unit	Advantages	Disadvantages
Sentence	Naturalness throughout	Usually impractical to prerecord every sentence that might be needed
Word	Naturalness within the word	Interword articulation may sound mechanical; usually impractical to prerecord all the words that might be needed
Syllable	Naturalness within the syllable	Good inter-syllable articulation is difficult to achieve, making individual words unintelligible; usually impractical to prerecord all the syllables that might be needed, though not as seriously as with words or sentences
Allophone	Naturalness within the allophone, especially vowels; much fewer units need be prerecorded than with other choices	Good articulation between most allophones is difficult to achieve, making individual words unintelligible; stop consonantal allophones are difficult to isolate for prerecording
Diphones	Naturalness on the allophonic level; much better articulation between diphones than between allophones	Consonant clusters do not always sound natural; syllable lengths do not always sound natural; impractical to extract and prerecord all possible allodiphones from naturally spoken speech
Demisyllables	Naturalness on the syllable level; natural sounding consonant clusters; natural sounding syllable lengths	Articulation at syllable boundaries may sound unnatural, making individual words unintelligible, though not as seriously as with syllables; impractical to extract and prerecord all possible allodemisyllables from naturally spoken speech

tion of prosodic characteristics on the reconstructed waveform, in particular, a smoothing of the transitions between concatenated units. Until recently computing power was insufficient to do this effectively.

There is a tradeoff between LPC and the various methods of waveform coding. With waveform coding the speech units themselves sound natural, but the transition between units is stilted. With LPC coding the transitions are smoother, but the speech has an inescapable mechanical quality that listeners find both unnatural and unpleasant.

In the early 1990s speech scientists developed an efficient method to manipulate the prosody of a reconstructed waveform. *Pitch synchronous overlap-add* (PSOLA) is a technique for changing the pitch, segment length and loudness (hence, the stress and intonation) of spoken units after waveform reconstruction. The technique may be applied to waveforms both in the time domain, where it is called *time domain PSOLA* (TD-PSOLA), and the frequency domain, where it is known as *frequency domain PSOLA* (FD-PSOLA).

TD-PSOLA is used to speed up or slow down selective portions of the speech, effectively controlling segment duration. The idea is similar to the one underlying DTW (Section 3.8.1). FD-PSOLA is used to adjust the pitch (fundamental frequency) by "warping" in the frequency domain, that is, by compressing or expanding the signal through the deletion or insertion of harmonics. (It is possible, but inefficient, to adjust pitch in the time domain using TD-PSOLA. The advantage is not having to apply the DFT to produce a frequency domain representation.) Loudness is adjusted in FD-PSOLA by modifying the amplitude.

It is safe to say that once properly instructed, speech synthesizers will be able to speak as intelligibly, and nearly as naturally, as an articulate human by the end of the 20th century. The catch is: *how do we properly instruct the machine?* That is the topic of the Section 4.4.

4.4 Text-to-Speech Processing

The language of speech synthesizers is not the language of the human user. Parametric synthesizers accept the name or symbol of a menu item such as [i] and construct the corresponding wave form. Concatenative synthesizers also accept symbolic input that determines what unit to concatenate next and controls the prosodic output. We humans, however, would prefer to tell the synthesizer what to say in ordinary written language. Moreover, many applications of speech synthesis, such as reading machines for the visually impaired, are forced to accept textual input.

The text-to-speech process is converting writing to the input symbols of the speech synthesizer. It is tantamount to what you do when you read aloud. To gain an appreciation for the difficulty of the task, hearken back to when you learned to read. If you don't remember, think about what a younger sibling, or a child you know, underwent in learning to read. It isn't easy, and English is especially difficult. If your personal experiences don't convince you, think of the magnitude of the effort, not to mention the expenditure, of teaching Johnny and Janie to read in the public schools. Figure 4.2, similar to Figure 4.1, shows a model of text-to-speech processing.

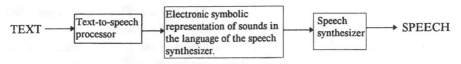

Figure 4.2 Model of a text-to-speech system.

4.4.1 Rule and Exceptions

Reading aloud is a complex mental activity. You must determine the pronunciation of each allophone, you must get the relative segment length, pitch and stress—together called **rhythm**—sounding right, and you must make the overall intonational contours sound natural. Whew! How do you manage to do all that so effortlessly? Let's examine what kinds of rules are followed and what the exceptions are, when converting text to speech.

There are rules of reading, though that fact is somewhat obscured by the overwhelming number of exceptions, illustrated in the following poem, entitled *English,* published long ago in a British newspaper.

> I take it you already *know*
> Of *tough* and *bough* and *cough* and *dough?*
> Some may stumble, but not *you,*
> On *hiccough, thorough, slough* and *through?*
> So now you are ready, perhaps,
> To learn of less familiar traps?
> Beware of *heard,* a dreadful *word*
> That looks like *beard* and sounds like *bird.*
> And *dead,* it's *said* like *bed,* not *bead;*
> For goodness' sake, don't call it *deed!*
> Watch out for *meat* and *great* and *threat.*
> (They rhyme with *suite* and *straight* and *debt.*)
> A *moth* is not a moth in *mother,*
> Nor *both* in *bother, broth* in *brother.*

Suppose you wanted your synthesizer to be able to read this poem correctly. Of course you could transliterate it by hand and feed the synthesizer diphone by diphone. You could work out the intonation and rhythm, and code that into symbols for controlling stress, pitch, segment duration and loudness. You could spend a week, full time, doing it, too! The sensible approach is to identify as many regularities as possible for pronouncing letters and their combinations. Then exceptions could be identified. Finally, you could take a stab at prosody.

Consider the first line. What are the regularities? *I* by itself (surrounded by white spaces) is pronounced [ai]. The letter *t* is pronounced [t^h] at the beginning of a one syllable word. (This implies that the text-to-speech scanner must look ahead and not simply proceed letter by letter.) The letter *a*, when followed by a consonant and the letter *e* at the end of the word is pronounced [ā]. The consonant *k,* when it occurs at the end of the syllable is pronounced [k] (unaspirated). The letter *i* is pronounced [i] when followed by a consonant at the end of a word. (It would be [ai] if followed by a consonant and an *e* at the end of the word.) The *t* of *it* will either be [t] or [t⁼] (unreleased *t*); *y* is pronounced [y] at the beginning of words.

The digraph *ou* is a problem. It has several pronunciations, four of them mentioned in the second line of the poem. Our job is to find as many rules for *ou* as possible; all other words containing *ou* will be exceptions. Given common words like *out, our, ours, hour, ouch,* we might propose that *ou* be pronounced [ou] in general. However, when followed by *-ght,* as in *bought, fought, sought, thought,* the pronunciation is [ô]; when followed by *-gh,* the pronunciation is [ŭ]; and when followed by *ld,* the pronunciation is [u].

Thus there are regularities, subregularities, sub-subregularities, and so on. The order in which these rules apply is important. The least regular rules apply first, then the more regular rules. The general rule is applied last. In the case of *ou* we first see if the *-ght* rule applies—it's the least regular (i.e., most restrictive) rule. If it does, the *ou* is assigned [ô] and we're done. If it doesn't, we check the rule for -ough; if it applies the *ou* is pronounced [ŭ], and we're done. Otherwise we check for *-ld,* and so on. All other occurrences of *ou* are pronounced as [ou].

Even with four rules there are plenty of exceptions: *bough, cough, through, thorough, dough, hiccough.* The exceptions can be viewed as the most restrictive possible "rules." They are placed in a database dictionary along with their pronunciation. They are checked first, before any other rules apply. For reference purposes, we number the rules in Table 4.5.

Table 4.5
Some Rules for the Pronunciation of *ou*

Rule Number	Context	Pronunciation
Exceptions (e.g., *dough*)	Whole words	Given in dictionary
1	*-ght*	[ô]
2	*-gh*	[ŭ]
3	*-ld*	[u]
4	elsewhere	[ou]

Let's run through the algorithm for a few common words. *You* is found in the exceptions dictionary; its pronunciation assigned; and that's that. *Rough* is not an exception. Rule 1 doesn't apply but rule 2 does. *Could, should,* and *would* are not exceptions, and neither rules 1 nor 2 apply, but rule 3 does. *Bough* is an exception, even though it follows the general rule. It's exceptional as an "ough" word. Many common words are not exceptions and not subject to the first three rules, such as a*bout, bound, doubt, foul, flour, hound, joust, mouse, noun, pouch.* Rule 4 handles them.

There are many exceptions: *boulder, cougar, nourish, pour, touch.* It may be possible to winnow out a few more subregularities in an attempt to reduce the number of exceptions, but one must always take care not to produce more exceptions than one is eliminating. A rule that says "pronounce *ou* as [ō] when preceded by a bilabial consonant (i.e., *b, m, p*) would cover *boulder, mould, moult, poultry, pour,* but it would make exceptions out of such common words as *bounce, mouse and pout,* so it's a bad idea.

Our prolonged discussion of *ou* illustrates the enormous scope of one aspect of the text-to-speech task: letter-to-allophone translation. However, letter interpretation is only part of the picture in converting text to intelligible, natural sounding speech.

4.4.2 Morphological Analysis

Many words can be broken down into **morphemes.** A morpheme is the smallest unit of linguistic meaning or function. For example, *baseball* contains two morphemes: *base* and *ball.* Those words themselves are each a single morpheme. *Baseballs* has three morphemes: the two just cited, plus *s.* The plural morpheme, spelled *s* or *es,* is a functional morpheme. Table 4.6 gives several words and their morphological decomposition.

Consider entries in the exceptions dictionary like *cough* and *dough.* To avoid the need for additional entries such as *coughs, coughing, coughed, doughier, doughiest, doughiness, doughboy, doughface,* we decompose words into their component morphemes and apply the letter(s) pronunciation rules to them. If *coughs* is encountered, morphological analysis breaks it into *cough + s. Cough* is found in the exceptions dictionary, and a simple rule says to pronounce *s* as [s] when preceded by a morpheme boundary and a voiceless consonant. (A similar rule would cause s to be pronounced [z] if the verb were *calls.*)

A second need for morphological analysis is seen in words such as *hothouse.* If this word is taken to be a single morpheme, the pronunciation of the *th* in the middle will be wrong. However, analyzed correctly as two morphemes, *hot + house,* the correct pronunciation is easily achieved. The pronunciation of *th* when it occurs between vowels is dependent on morphemic analysis. When

Table 4.6
Some Words and Their Component Morphemes

Word	Morphemes
Run	Run
Runs	Run + s
Homeruns	Home + run + s
Cap	Cap
Uncap	Un + cap
Uncapped	Un + cap + ed
Uncapping	Un + cap + ing
Hothouse	Hot + house
Nonentity	Non + entity
Photography	Photo + graph + y
Vaccine	Vaccine
Vaccinate	Vaccin + ate
Longer	Long + er
Shortest	Short + est
Redoable	Re + do + able

th occurs between vowel letters within a single morpheme, it is usually pronounced [dh], as in *bother, brother, father, mother, other, either, seethe, bathe,* etc. (*Ether* is an exception.) Between vowels where one vowel is in one morpheme and the other in a different morpheme, the *th* is pronounced as it would be in its own morpheme: *bethought, rethink, seething.*

Morphological analysis is easier said than done. Many polymorphemic words undergo spelling changes that disguise their root morphemes. Table 4.7 offers a few of the many hundreds of challenging decompositions. Bear in mind that even when the morphemes are properly identified, the difficulties of determining the correct pronunciation are still present.

A great deal of ingenuity goes into the programming of automatic morpheme decomposition, made even more difficult by the constraint of operating in real time. Nonetheless, modern speech synthesizers do a capable job of it.

4.4.3 Articulation Effects

Naturally spoken human speech is mostly all run together, except for breath pauses. Speech sounds are often shortened to the point of omission (the second *e* in *general*), and adjacent speech sounds frequently overlap and blend into

Table 4.7
Some Words and Their Component Morphemes With Spelling Changes

Word	Morphemes
Democracy	Democrat + y
Judgment	Judge + ment
Desirable	Desire + able
Beneficial	Benefit + ial
Description	Describe + tion
Canadian	Canada + ian
Economic	Economy + ic
Galactic	Galaxy + ic
Profundity	Profound + ity
Papal	Pope + al
Assumption	Assume + tion
Circulate	Circle + ate
Satisfaction	Satisfy + tion
Remission	Remit + sion
Submissive	Submit + ive
Excessive	Exceed + ive
Decision	Decide + sion
Conception	Conceive + tion

something different from either individual sound (the *t* and *y* of *got you* when pronounced "gotcha"). It's this "sloppiness" that makes speech sound natural to the human ear. Whether analyzing or synthesizing speech, one rarely knows at what point one speech sound blends into the next.

Morphemic analysis creates more speech units to be fit together than word analysis. This in turn creates more boundaries between units where articulation effects must be figured out. If *hypocrisy* is to be analyzed as *hypocrite* + *y*, then during text-to-speech it must be determined that the *t* is to be pronounced [s], not [t] as the morphological analysis suggests, nor [z] which is possible based on the spelling (as in *business, nosy, advisory*). Then the conversion to diphone notation (or whatever) can take place. As Table 4.7 suggests, the number and complexity of these rules is daunting.

A partial morphological analysis seeks to decompose only words containing certain function morphemes like *s, es, ed, ing, est,* etc. Thus *hypocrites* will be decomposed into *hypocrite* + *s,* but *hypocrisy* will have its own entry and be treated as if it were monomorphemic, though technically its *hypocrite* + *y.*

The need to manage articulation effects between words remains. A natural pronunciation of *Are you on your feet yet?* will combine the *n* of *on* with the *y* of *your* into a palatalized nasal, the "n" of Spanish *señor.* Also, as we've seen, the *t-y* connecting *feet* to *yet* will be pronounced [ch]. All of the words are connected, as if they were spelled *areyouonyourfeetyet,* which is much closer to the representation sent to the synthesizer than text with white spaces.

Failure to properly determine articulation effects at any level—word, morpheme, submorphemic unit—will lead to speech that sounds mechanical and unnatural.

4.4.4 Prosody

Today's synthesizers are capable of all the nuances of speech necessary to natural-sounding prosody. The problem lies in knowing how to instruct the synthesizer. The goal of producing synthetic speech with natural prosody is a text-to-speech challenge.

Poor prosody is the major cause of unnatural sounding synthetic speech. On the allophone level, prosody encompasses loudness, pitch, and duration. On the syllable and word level, prosody refers to stress, which is a secondary effect of changes in loudness, pitch, and duration. On the phrase and sentence level, prosody refers to **rhythm** (or **meter**) and intonation. Rhythm is the overall effect of stress placement within a phrase or sentence. Although most people think of rhythm as associated with poetry—*ONCE upON a MIDnight DREARy, AS i PONdered WEAK and WEARy*—all spoken utterances have a certain inherent rhythm that is vital to the intelligibility and naturalness of speech. Intonation is imposed over, and blended with, the pitch and stress patterns of individual words and phrases. It is the continually changing pitch contours of entire phrases and sentences. The production of naturally sounding speech is no less complicated than the performance of a symphony—attention must be paid to the details of rhythm and melody.

4.4.4.1 Stress

Assuming morphemic analysis, one first deals with the placement of stress within the morpheme. There is an endless amount of literature and speculation regarding the rules of stress placement within morphemes, let alone within words. (See [1] for a classical treatment of the subject). The rules are complex and rarely without exception. Inviolable generalizations appear not to exist.

To achieve a grasp of the difficulty of discovering what the stress rules are, consider the monomorphemic word *man.* By itself in a sentence, as in *He is the man I love,* it receives primary stress. But as the second morpheme in *gentleman,* it is unstressed. Primary stress is on *gen,* and *man* is pronounced [mən], proof that the syllable is unstressed. In the sentence *I love the man, love* receives primary stress and *man* receives secondary stress. (Say it out loud.) If *man* were to have primary stress, it would be interpreted as contrastive, as if I were saying *I love the MAN* (not the woman).

Now consider the morpheme *photo.* In the sentence *That is a photo of me,* the second syllable, *to,* receives secondary stress. We know that because the stress on *pho* is stronger, but the full vowel value is pronounced in *to.* However, when we combine the morphemes *photo* and *graph* to produce the word *photograph,* the syllable *to* is destressed entirely, and pronounced [tə]. A pronunciation of photograph with the *to* pronounced [tō] will sound robotic.

Add the morpheme *-y* to form *photography* and the stress rearranges itself. Now primary stress is on the second syllable *to,* secondary stress is on the final syllable *phy,* and the odd numbered syllables are destressed and their vowels reduced to schwa. Any violation of this pattern will sound abnormal. Even this analysis is oversimplified. The observant reader will note that the vowel quality in the syllable *to* of *photography* is different than it is in *photo.* It's [o] in *photography, but* [ō] in *photo.* There are many such examples in English, for example, *telephone - telephonic - telephony,* where primary stress is on a different syllable in each word.

Some of these stress patterns are predictable and rule-governed. Some are idiosyncratic. Others are determined from syntactic or semantic considerations. Many words in English differ in meaning depending on which syllable is stressed. In *He is content to stay,* primary stress is on the second syllable of *content.* In *Its content is vulgar,* primary stress is on the first syllable.

Similar cases span several words. The noun phrases *hot dog* and *hotdog* differ only in stress placement. (Remember, the white space isn't present when speaking.) The overheated canine has primary stress on *dog;* the foodstuff has primary stress on *hot.* Similarly for differences between *red coat* and *Redcoat, white house* and *White House,* etc. The words *light - house - keeper* combine to form expressions of different meaning depending on stress placement. With primary stress on *light,* it means "keeper of a lighthouse." With primary stress on *house,* it means "housekeeper who does light house work." Again, there are dozens of similar combinations such as *grand - father - figure,* or *black - board - eraser.*

If you are a native speaker of English, or a very accomplished speaker of English as your second (or third, etc.) language, you intuitively know these facts of stress placement. You produce them in your own speech, and comprehend them when listening, and when they are not correct, you may be confused or misled. You have this knowledge because you know the syntax and semantics of English, and you know the rules of stress placement, subconscious though this knowledge may be. Programming this knowledge into a speech synthesizer is a significant challenge, but a failure to do so will result in synthetic speech that sounds, well, synthetic.

Speech scientists have two basic methods for determining correct stress placement. The first is to analyze sentences syntactically and semantically and apply known stress rules. An analysis of *its content is vulgar* will determine that *content* is a noun and therefore receives syllable initial stress. A similar analysis of *he is content to stay* reveals that *content* is an adjective and receives syllable final stress. The same tactic works on *red coat* versus *Redcoat*. The former is an adjective + noun combination of separate words and takes primary stress on *coat,* and secondary stress on *red. Redcoat* is a compound noun—one word formed by concatenating an adjective and noun—and therefore receives primary stress on *Red* and secondary stress on *coat.*

The second method is to apply usage statistics. Enormous databases of sentences are collected, both from written and spoken sources. The occurrences of *CONtent* and *conTENT* within various contexts are tabulated. The computer decides how to pronounce *content* on the basis of those statistics, a process of "intelligent guessing." Both approaches to disambiguating words like *content* only became feasible in the 1990s with the advent of computers with fast processors and vast storage capabilities at a relatively low cost.

4.4.4.2 Intonation

English is an intonational language. Naturally spoken English is melodious. The fundamental frequency of speech, established by the vibrating vocal cords, varies throughout the sentence. Moreover, this pattern of intonation affects meaning. *He ate meat,* spoken with falling intonation, is interpreted as a statement; the same sentence with rising intonation is interpreted as a question; and with dramatically rising intonation, is interpreted as surprise or outrage.

Overall intonational patterns play a role in disambiguating sentences that contain the same words. For example, *What's that in the road ahead?* versus *What's that in the road, a head?* where the comma suggests the different intonational patterns. A somewhat more subtle ambiguity is found in *I left directions for you to follow,* as discussed in Section 1.2.4.

Intonation imposes itself over the pitch and stress patterns of individual words. Consider *She is an elevator operator.* Both *elevator* and *operator* have primary stress and highest pitch on their first syllables, The pitch within each word falls for the second syllable, rises for the third but not to its previous height, then falls off, as illustrated in Figure 4.3.

In the statement, *She's an elevator operator,* the two words have the same relative pitch contour internally, but the overall falling intonation causes the ab-

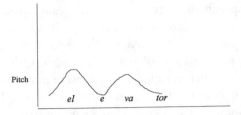

Figure 4.3 Pitch contour for *elevator* and *operator.*

solute pitch levels to decrease. If it were a question, the absolute pitch level in *elevator* would be the same, but the intonational pattern of a question would force pitch levels to rise in *operator.* This is illustrated in Figure 4.4.

Deviations from these patterns evoke all kinds of reactions from people. Some years back a student of mine did a term project cataloging responses of people to intonationally abnormal speech. There were several subjects who

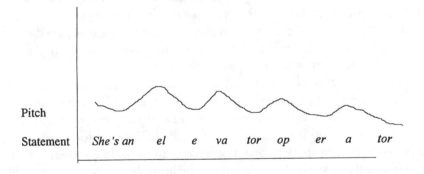

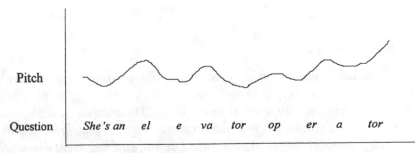

Figure 4.4 Pitch contours for the statement and question: *She's an elevator operator.*

thought it was the speech of native speakers of Scandinavian or Asian languages. A surprising number (about 20%) thought it was synthetic speech—that may be an artifact of doing the survey in the vicinity of a university, or evidence for the rising pervasiveness of synthetic speech. Others perceived the speech as "singsong." Two persons of the 50 that were interviewed thought the speech was poetry. (There was nothing poetical about the sample sentences that I could see, but the "logic" is clear: Poetry has a rhythm that affects intonation; therefore any speech with affected intonation is poetry.) About 10% claimed that didn't notice anything unusual.

You have undoubtedly experienced difficulty understanding the English of a nonnative speaker, or even of someone who doesn't speak your dialect. When you listen analytically you often realize that the person is pronouncing all the sounds correctly—leastwise, sufficiently correctly—but the rhythm diverges from what is expected. Native speakers of French, a language without the prominent stress patterns of English, often speak English in a rather stressless way, because that is how French is spoken. That can make them hard to understand. (Conversely, English speakers who speak French are equally difficult for native French speakers to understand because they try to impose stress patterns where none exist.) Native speakers of some of the languages of the Indian subcontinent, many of whom speak English as a second language, are difficult for Europeans or North Americans to understand solely because they fail to conform to the rhythm of those English dialects.

To ensure the intelligibility and naturalness of speech synthesizers, we must make every effort to make rhythm and intonation conform to those of native speakers. An incorrect stress pattern is worse than no stress pattern at all insofar as comprehension is concerned, so if in doubt, it's best for a synthesizer (or a person trying to speak a foreign language) to speak in a monotone with each syllable equally stressed. The effect will be horribly dreary, but intelligibility will suffer least. To achieve naturalness in this realm the text-to-speech process must be well-informed about the stress pattern of morphemes, how those stress patterns change in words, and how the syntax and semantics of the utterance further impacts these prosodic parameters.

4.4.5 Special Problems

Many special cases arise in text-to-speech processing by computer. We have already encountered the exceptions dictionary, where thousands of words that do not follow the letter-to-allophone rules are listed. The exceptions are not necessarily rarely used words. Many common words, such as *put, been, does* have exceptional pronunciations (compare with: *but, cut, gut, hut; keen, queen, seen, teen; foes, goes, hoes, toes*). Many other kinds of "words" offer special pronunciation problems. In this section we consider some, but certainly not all of them. To make this section more concrete, imagine that we want a text-to-speech system capable of reading novels, newspapers, and electronic mail, but let's not worry about textbooks, magazines, and the like, where difficulties arise such as speaking a chemical formula or mathematical chart. Let's also exclude for now the unique challenges of conveying the contents of a computer screen by voice. All these problems, and more, are currently subjects of research, some of which will be discussed in Chapter 7.

4.4.5.1 Heteronyms

Heteronyms are different words that are spelled the same but pronounced differently, as in *He will lead us to the lead mine.* They are problematic for text-to-speech because the correct pronunciation is determined by the syntax and semantics of the context, just as it was in certain cases of stress placement.

There are over 300 heteronyms in English, many of them common words such as *dove, does, live, sow, wind, bass.* At one time, the text-to-speech strategy was to use the more common word. Thus *does* would be pronounced [dŭz] in all cases, and if the text was *Does eat oats,* tough luck.

The "most common use" principle is the simplest statistical strategy available. More sophisticated statistics may be garnered by examining large amounts of language data and observing likely contexts for the two words. The noun *does* occurs more frequently after *the* than the verb *does;* and the verb *does* occurs more frequently before *not* than the noun. A question arises as to how much context to examine. One could look at surrounding words or preceding words. Another question is how many surrounding or preceding words. A popular strategy is to look at **trigrams,** which are word triples. The word of interest is the third word, the preceding two words are the context.

One drawback to statistical brute force is just that: brute force consumes a lot of computational resources. Another drawback is that it is not 100 percent accurate. In a trigram analysis, *does* is doomed to be mispronounced in either *We saw the men shooting does not bucks* or *Seeing the men shooting does not bother us,* because the trigram context *men shooting* is the same.

There is little doubt that the best strategy would be a syntactic and semantic analysis of the context in which a heteronym occurs. That's essentially what people do (subconsciously) when reading aloud. The surrounding words alone, without regard to their syntactic structure, may not provide sufficient information. Consider the pronunciation of *does* in the following sentences:

The cowardly hunters may shoot does not bucks.
Knowing that the cowardly hunters may shoot does not scare John.

It is syntactic structure together with the word context that determines that *does* is the plural of the noun *doe* in the first sentence, and the third person singular verb form of *do* in the second sentence.[2]

Semantic analysis is confined to finding words in the context with meanings related to the heteronym in question. For example, one might look for references to fishing or to music in attempting to disambiguate *bass*.

Many names are heteronymic, such as *Reagan*. In such a case, there is little point to examining syntax and semantics because proper names are always nouns and usually don't have any intrinsic meaning. The best strategy is a statistical one: *Reagan* is far more often pronounced [rēgən] than [rāgən]. A text-to-speech program that changed with the times might check context for references to political history, but such attention to minutia is in the future.

4.4.5.2 Proper Names

Open the white pages of the phone book at random. Begin reading the different last names. How far do you think you'll get before you reach a name you cannot pronounce correctly with confidence? I don't mean totally clueless, as you might be with *Hryshchyshyn* or *Hsieh*. I mean any case where you are unsure because two or more possible pronunciations come to mind. The correct pronunciation of deceptively simple names like *Hudak* ([hūdăk], [hŭdək], etc.?) or *Smythe* ([smith] or [smaith]?) cannot always be ascertained. I tried it myself a few times and could rarely get through ten surnames.

Proper names show up everywhere, from *War and Peace* (there are hundreds) to newspaper headlines. Their pronunciation is the bane of radio and television announcers, and nowadays, of text-to-speech synthesizers. The problem is simple enough to state. Many proper names do not follow the regular rules of letter-to-allophone conversion. There are two reasons. They are exceptional, the way *put* is exceptional, or they follow a different set of rules.

With today's computers it is technically feasible to enter every known proper name and its pronunciation into a database that is like an exceptions

---

2. It's not possible to go deeply into syntactic theory in a book of this kind, but readers with some background might observe that the first sentence has the structure noun phrase-auxiliary verb-verb phrase. The noun phrase is *the cowardly hunters;* the auxiliary verb is *may;* and the verb phrase consists of the verb *shoot* followed by the noun phrase *does not bucks*. The second sentence has the basic structure of noun phrase-verb phrase, where the noun phrase contains the sentence remnants *knowing that the cowardly hunters may shoot*. The verb phrase consists of the verbal complex *does not scare* and the noun phrase *John*. If this footnote makes no sense to you, you needn't worry, although you might muse over the fact that this analysis, or something similar to it, is representative of every English speaker's subconscious knowledge of the language.

dictionary. However, it is not possible in practice to include every proper name for several reasons. First, many such names are unique (about one third of last names in the U. S. Social Security files occur once). There is little opportunity to discover them and determine their correct pronunciation for entry into the database. Another reason is that many proper names are of foreign origin and it is difficult to anticipate when they will become relevant for an English text-to-speech process. Yet another reason is proper names are being continually created, such as the names of newly emergent African nations. A current vogue among many Americans is to give their newborn a unique name.

If a name is not found in the dictionary, the synthesizer attempts to pronounce it using rules. The spelling rules we have seen in this chapter are Anglo-Saxon in origin and will work on many Anglo-Saxon proper names such as *Norwoodsman*. When they don't, as in *Nowlan* ([nōlən] or [noulən]), there is little recourse and mistakes will be made. When an unknown proper name is encountered, someone can ascertain the pronunciation and add the word to the dictionary of names.

If a name is not of Anglo-Saxon origin, it is unlikely that the spelling rules will work. For example *ch* is typically pronounced [ch] in English, [sh] in French, [k] in Italian, and as a velar fricative in German *Bach*. A technique in use today is to examine a name for its language of origin, and then apply a set of rules appropriate to that language. For example, a polysyllabic name ending in *ier* is most likely French; if it ends in *etti* it is most likely Italian. Using those clues, a name such as *Chabrier* can be identified as French and pronounced correctly as [shabrēā].

There are great challenges in pronouncing names from countries such as Poland, Latvia, Russia, Finland, etc. Regularities are evident in the spelling of these names, so the rule-based approach works to a degree. For example, words that begin with *sz* are most likely of Polish origin, and the *sz* is pronounced [sh]. On the other hand, all languages have their idiosyncratic spellings, for any number of reasons. Basketball fans may long wonder why Duke University's coach Mike Krzyzewski pronounces his last name [shəshevski]. Fortunately, he is sufficiently well-known so that when synthesizer dictionaries of proper surnames are created, his surname name will undoubtedly be listed.

When we get away from names of European origin the difficulties mount. Nonetheless, there are regularities in the pronunciation of names of Chinese, Southeast Asian, South Asian, and African origin. Moreover, many such names have signature combinations of letters that reveal their origin, such as word initial *hs* for Chinese, *bh* for many South Asian languages, *ng* for Vietnamese and some African languages [2].

Considerable research into name pronunciation takes place at Bellcore (Bell Communications Research). Their speech synthesis system, called ORA-TOR, provides the reverse-directory service currently available in the Illinois region [3].

4.4.5.3 Foreign Words in English Texts

Before I learned French, I knew perfectly well how to pronounce *mariage de convenance, ensemble, and rouge:* I anglicized them, which means to mispronounce them in the English manner. After a year of college French I could say them right, but when I did it sounded snobbish. Being persistent, I learned just the right degree of mispronunciation to show my snobby friends I knew some French and my Anglo-Saxon pals that I wasn't above substituting a few hearty English allophones where they didn't belong.

A similar conundrum faces the designer of text-to-speech systems for use in English speaking countries. Should the *g* in *mariage* be pronounced [zh] as in good French, or [j] as in good English? Should the *e* in *de* be pronounced [ā] (as in good Spanish, which pronunciation good English has adopted), or [ə] as in proper French? Are the vowels in *ensemble* to be fully nasalized, and the *n* and *m* omitted entirely, as in French; or partially nasalized with the nasal consonants pronounced, as if it were English?

Many borrowed words such as *ensemble* have become so much a part of English that they are given an official pronunciation in dictionaries. The text-to-speech algorithm need only follow the dictionary's recommended pronunciation. For foreign words that aren't in a dictionary of American English, the same strategy used on foreign names applies, which is to try to identify the language source and apply spelling rules for that language. Many such words are to be found in the menus of ethnic restaurants. Others are *la dolce vita, sans souci, uhuru, weltgeist, amor vincet omnia.* The synthesizer, like the person who doesn't want to seem snobbish, must carefully choose a "mispronunciation" that will maximize the listener's comprehension, even if it means enunciating the foreign word with "an English accent."

4.4.5.4 Abbreviations

Most texts contain a wide variety of abbreviations, where I take the term to include numbers, dates, times-of-day, monetary expressions, acronyms, special symbols such as #, and punctuation, as well as traditional abbreviations like *Mr.*

Written English uses thousands of abbreviations. Many are standard. Table 4.8 offers a smattering of abbreviations and an indication of how to pronounce them.

Table 4.8
Some Abbreviations and Their Pronunciation(s)

Abbreviation	Spoken As
Mrs.	Missus
Dr.	Doctor, drive
Ph.D.	Pee aitch dee
St.	Street, saint, stanza
Ch.	Chapter, chaplain
1,111	One thousand, one hundred (and) eleven
111th	One hundred (and) eleventh
44/100	Forty-four hundredths; forty four one hundredths
1.11	One point eleven; one point one one
5/19/40	May nineteenth, nineteen forty; the nineteenth of May, nineteen forty
5:30 A.M.	Five thirty a em
(*#)	Left parenthesis asterisk pound sign right parenthesis
$44.44	Forty-four dollars and forty four cents
NASA (National Aeronautics and Space Administration)	Nassa
NCSU (North Carolina State University)	En see ess you
DOD (Department of Defense)	Dee oh dee

Common abbreviations may be put in a dictionary. Ambiguities are resolved in the usual way: by context or frequency of occurrence. Using context it is easy to disambiguate *Dr. Einstein lives on Riverside Dr.* Using statistics one would choose to pronounce the abbreviation *Ch. as chapter,* that being a more common usage than *chaplain.* Of course if the *Ch* is followed by a numeral of any kind it most likely abbreviates *Chapter.*

A good text-to-speech program handles numbers (both cardinal and ordinal), fractional expressions, decimal numbers, dates, times of day, currency amounts, and punctuation. The period and comma are represented by pauses of varying lengths. The colon and semicolon engender somewhat shorter pauses. The question mark produces rising intonation. Certain subtleties such as the special intonation of "scare quotes" (the previous quotation marks are an example) are not usually expressed, probably because synthesizers are not equipped with fingers for inscribing quotation marks in the air.

Other special symbols such as @,#,%,&,*,(,),-,—,/,<,>,[,] are generally spoken as "at symbol," "pound sign," "percent," "ampersand," "asterisk," "left parenthesis," "right parenthesis," "hyphen," "dash," "slash," "left angle bracket," "right angle bracket," "left square bracket," and "right square bracket." The list is not exhaustive as a brief perusal of a typical keyboard will reveal.

There are thousands of acronyms in use today. They pose an interesting problem for text-to-speech synthesis: When do you pronounce them as if they spelled a word, such as [năsə] for *NASA;* and when do you read the letters, as in *DOD* (dee oh dee—Department of Defense)? If the letters do not conform to the spelling of an English word, the acronym is most often spelled out, as in *NCSU.* Sometimes a vowel is inserted that allows a slightly misleading pronunciation such as *rifra* for *RFRA* (religious freedom restoration act). However, even when an acronym is perfectly pronounceable, such as *DOE* (Department of Energy), or *UNC* (University of North Carolina) it may still be read off letter for letter, depending on custom.

Common acronyms like *FBI* can be stored in a dictionary along with their pronunciation. However, new ones appear in print from time to time and a strategy for speaking them is needed. The best strategy I know is to attempt to pronounce all acronyms as words if they contain four or more letters, as *you ness coe* for *UNESCO.* If the acronym contains three or fewer letters, or if the letter combination is unpronounceable, then reel off the individual letters one at a time.

4.5 Concept-to-Speech

In text-to-speech processing, the computer input is a written text intended to be read as written. In concept-to-speech processing, the computer input is in the form of data, which the computer itself must convert to a written text to be read. A model of concept-to-speech processing is shown in Figure 4.5, similar to Figures 4.1 and 4.2.

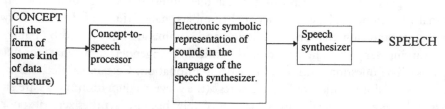

Figure 4.5 Model of a concept-to-speech system.

A principle goal of computer science is a machine with which a human can have an intelligent conversation. The effort to achieve this goal has spanned nearly the entire history of computers. Its roots lie in the Turing test discussed in Section 3.2. The "test" requires an interrogator to converse with a computer trying to pass itself off as a human. To accomplish this feat the computer must express its own "ideas" in speech.

Assuming that "ideas," or "concepts," can be represented in a computer, and that those thoughts are prelinguistic—that is, not already expressed in human language—we would like to know how the computer chooses the appropriate words and sentences. This is a topic on which whole books are written [4], so our discussion is necessarily limited.

Computer "thoughts" or concepts are mostly either assertions or questions. Assertions may indicate an affirmation or denial of some fact, a statement of some fact, an understanding of some fact, an acknowledgment of some action, a request for someone to do something, etc. Questions may be asking for the location of an object, asking if a certain action should be done, asking how to do a certain action, requesting that a certain action be done (a polite form of commanding), etc.

If the computer wants to respond to a question such as *Is two plus two equal to four?* it first determines whether it agrees. If it does if formulates its response first in the abstract, for example ASSERT(AFFIRMATIVE). The language generator takes this as input, and outputs a "text" such as *Yes, it is.* The text-to-speech algorithm completes the concept-to-speech process by having the synthesizer speak "Yes, it is." ASSERT(NEGATIVE) would engender "No, it isn't."

A more complicated situation might be asserting a certain measurement, say the distance between New York and Los Angeles. The computer's concept may be symbolized as ASSERT(STATE(MEASURE(DISTANCE(NEW YORK, LOS ANGELES). STATE indicates that the output is to be a statement of fact. MEASURE indicates that the statement of fact is a measurement. DISTANCE means that the measurement is a distance. NEW YORK and LOS ANGELES are proper names whose pronunciation is known to the computer. The language generator knows that a measurement has the sentential form THING-MEASURED *is* ACTUAL-MEASUREMENT. It is able to determine that THING-MEASURED is to be rendered as "the distance between New York and Los Angeles." It presumably knows the distance in miles, which replaces ACTUAL-MEASUREMENT. Finally, rules of syntax determine that the two parts are connected by the verb *is*. The final output is "The distance between New York and Los Angeles is two thousand, four hundred and sixty one miles."

A different kind of concept-to-speech is having the synthesizer deliver an oral report based on a database of facts. Databases such as daily sports scores, stock market results, weather reports, etc. are all possible sources of input. The computer conceptualizes the raw data, generates the appropriate natural language report, and reads it aloud. We will return to such applications in Chapter 7.

4.6 Languages of the World

They spell it Vinci and pronounce it Vinchy; foreigners always spell better than they pronounce.

Mark Twain, *The Innocents Abroad* (1869)

The examples in this book are naturally taken from English. After all, that is the language of the book. Nonetheless, I have striven not to be excessively Anglocentric with illustrations from many foreign languages, and in that spirit let me make some further remarks regarding the problems of synthesizing languages other than English.

We already observed that from a purely electronic standpoint, synthesizers are capable of producing speech sounds from any human language. The difficulties fall into two classes: prosodics and text-to-speech. Every language has its own prosodic pattern. Deviation from that pattern by a synthesizer will result in lowered naturalness and intelligibility. English, as we've seen, is a "stress" language. Different syllables in a word receive different stress, following a complex pattern that defies an algorithmic solution. Moreover, vowels in unstressed syllables in English are generally pronounced as a schwa, or elided altogether. Failure to observe this pattern results in synthetic speech that sounds, well, synthetic.

Many European languages, for example French, are "equal stress" languages and from that point of view are easier to synthesize, but other difficulties crop up. Naturally sounding French makes use of "liaison" and "elision," in which words are run together in a special way. In liaison, the final sound of a preceding word is pronounced as the initial sound of the following word, so that *petit ami* ("little friend") is pronounced as if it were spelled *peti tami*. On the other hand, in *petit chien* ("little dog") the final *t* of *petit* is not pronounced. In elision, certain sounds are omitted, so that the two syllables *le eau* ("the water") are pronounced as the one syllable *l'eau*.

Tone languages such as Mandarin Chinese and Thai, discussed in Chapter 1, require the synthesizer to pronounce syllables with the proper relative pitch in order to achieve good intelligibility and naturalness. Syllables in languages such as Japanese are to be pronounced of equal duration for naturalness. As well, Japanese, while not strictly a tone language, nonetheless has both tonal and intonational patterns that are as difficult for the synthesizer to get right as in English.

For some languages text-to-speech is a relatively straightforward process. Both Spanish and German, as examples, have relatively consistent spelling-to-phoneme rules. English, as we've seen, and other European languages are not as consistent. However, the real difficulties arise with languages such as Arabic or Hebrew, where only consonants are represented in the text, the vowels being filled in by speakers who know the language. Other problematic languages are Chinese and Japanese, when written using ideographic characters. Again deep knowledge of the language is needed to know how to pronounce a certain character, not to mention the problem of identifying the character from among the many thousands that are used to write the language.

4.6.1 Dialects

Under some circumstances it may be desirable to have a synthesizer speak in the dialect of the majority of listeners, not necessarily the standard dialect. One would not want to use an American English synthesizer in the United Kingdom, Ireland, or Australia, and conversely, nor would a Yankee-sounding speech synthesizer be desirable in an application in rural Alabama. Indeed, while the languages of the world number about 5,000, there are tens of thousands of dialects.

When different dialects are spoken in different countries, a synthesizer is generally programmed to speak the standard dialect of the country where it is designed—broadcast English in the United States, received pronunciation in the United Kingdom, Parisian French in France, and so on. Within a country, if an application is judged to be more effective when speaking the local dialect, then the software can be customized to fit that need. This is a two-step process. First, the synthesizer's rendering of particular allophones may be subtly different and require adjustment of synthesizer parameters such as those in Table 4.1. Second, the text-to-speech process would have to be changed to accommodate the idiosyncrasies of the dialect.

4.7 Performance Evaluation

Speech synthesizers may be evaluated along three dimensions: intelligibility, comprehensibility, and pleasantness/naturalness. These are not the only criteria, but they are fundamental. Other factors in measuring the worth of a synthesizer are ease of use, response time, quality of user documentation, effectiveness of the text-to-speech processor (often, but not always, part of the synthesizer package), interface ability, and cost. (See [5] for details regarding material in this section).

Most of the evaluation procedures require humans to listen to speech synthesizers and respond to what they hear in a way that reveals how well they heard it. The results are statistical in nature and subject to the vagaries of human performance. Nonetheless, the human being is necessarily the ultimate judge of speech synthesis.

4.7.1 Intelligibility

Intelligibility is the ability of the listener to distinguish individual sounds. One way to measure it is to ask humans to take dictation from the speech synthesizer, which speaks a predetermined text. The dictation and the original text are compared and the errors tabulated to give a figure of merit. However, if the spoken text is grammatical, sensible English, then the listener may apply linguistic knowledge to determine what was said, leaving open the possibility that certain sounds were not actually intelligible.

A way around this problem is to present the listener with nonsense sentences. These may consist of actual words used in a semantically anomalous way such as *the six subjunctive crumbs twitch* from a poem by e.e. cummings; or they may consist primarily of nonsense words. Nonsense words are possible words in the language—words that comply with the phonotactic rules—that through chance are not actual words. The opening of *Jabberwocky* by Lewis Carroll is a prime example of such nonsense:

> 'Twas brillig and the slithy toves
>   Did gyre and gimble in the wabe;
> All mimsy were the borogoves,
>   And the mome raths outgrabe

Performance scores decrease dramatically when nonsense sentences are used in testing, but when comparing several synthesizers, the rank order remains the same.

4.7.1.1 Rhyme Tests

The earliest attempts to evaluate speech synthesizers systematically used rhyme tests. They were multiple choice, where one had to choose the actual word spoken by the synthesizer from among several rhyming words. For example the synthesizer would speak one of the six words *beet, feet, heat, neat, seat, wheat,* and the test subject would have to state which one.

A later version, called the *modified rhyme test* (MRT), used both rhyming sets and sets with differing final sounds, such as *seed, seek, seal, seam, seen, seep.* The MRT was a good test for the intelligibility of initial and final consonants but failed to test adequately the perception of consonant clusters such as *str-* or *-rst*. It also did not address vowel intelligibility, which it might have done with sets like *seep, sip, sap, soup, soap, sop.*

The MRT is a closed-response test. The answers are limited to a small list of words. In an open-response test the subject is asked to write down whatever word is spoken. This allows the testing procedure to winnow out weaknesses of the synthesizer, for example, the inability to produce intelligible word final clusters such as -rsts. Needless to say the open-response scores are lower than the closed-response scores, but as with nonsense words, rank order among synthesizers is unaffected.

Performance evaluation serves two purposes. One is to compare synthesizers with one another. The other is to compare synthesizers with human speech. To establish a baseline, the MRT and its variants are administered using clearly articulated human speech. The accuracy rate generally exceeds 99 percent for both open- and closed-response testing. Presumably this is the highest level achievable by synthetic speech.

4.7.2 Comprehensibility

The MRT sufficed to evaluate the intelligibility of individual sounds and words. More to the point, however, is the comprehensibility of phrases and sentences. For the most part, synthesizers do not speak in one-word utterances. Moreover, interword articulation effects, stress placement, and intonation are all sentence-level phenomena and cannot be effectively evaluated one word at a time.

The basic idea in evaluating comprehensibility is for a human to listen to some synthetic (or human) speech and then answer questions about it that reflect the degree of understanding. It is possible to comprehend entirely a passage in which not every sound or word is intelligible.

There are numerous variations on the basic idea. We'll discuss two of them. The first is to present listeners with short statements that are obviously true or obviously false and ask them to indicate which is which. Typical of these sentences are *Doctors prescribe medicine, Skyscrapers are short, Leopards have spots, Dishes can talk, When peeling onions people often cry,* and *A zip code has five letters.*

Three measurements are taken: accuracy, response time, and degree of certainty. Accuracy is simply the percent of correct answers. Though "obvious," a poor oral presentation may mislead the listener. In experiments carried out in my own laboratory, we often observed less than 100 percent accuracy.

Response time, sometimes called **latency,** is the amount of time taken to answer the question. The underlying assumption is the better the comprehensibility, the shorter the response time. This is borne out by experiments in which natural speech consistently has smaller mean response times than synthetic speech.

Degree of certainty measures confidence in the answer given. The higher the confidence, the more comprehensible the speech, presumably. Typically, subjects are asked to rate their confidence in each answer on a scale of 1–9, where "1" is a complete guess and "9" is absolute certainty.

The second variation on the basic idea will be hauntingly familiar to many readers because it is based on SAT (scholastic aptitude test) methods of measuring reading aptitude. A passage is read aloud in synthetic speech, after which questions are posed on the topic of the passage. The passages attempt to address obscure topics, such as the mating habits of the cockney sparrow, in order to prevent subjects from using their general knowledge to answer the questions.

The experiments described above can be used to rank different synthesizers and/or different types of speech, both natural and synthetic. One can also use the techniques to measure differences between native speakers and fluent, nonnative speakers in comprehending synthetic or natural speech.

Most evaluation experiments use subjects who are familiar with synthetic speech. Presumably, a person hearing synthetic speech for the first time will comprehend less, all things being equal, than a person who has had some exposure. That hypotheses was tested at North Carolina State University in the context of synthetic speech over telephone lines [6]. Listeners with as little exposure as three short sentences performed dramatically better on further synthetic speech than first-time listeners.

4.7.3 Pleasantness/Naturalness

The most subjective evaluation criteria are degree of pleasantness and degree of naturalness. The two are slightly different. It is possible for a voice to sound natural—that is, human-like—and be unpleasant; and it is possible for a voice to be judged pleasant and still have a machine-like quality, which is the case for modern, top-of-the-line speech synthesizers.

The most basic evaluation procedure is no more complicated than asking users to rate speech on a nine-point scale for pleasantness and naturalness. A typical questionnaire would state: "For each of the following questions, select the number from 1 to 9 that most closely corresponds to your rating of the voice you just heard: 1) How pleasant was this voice? 2) How natural was this voice?"

A more systematic method of evaluating naturalness is presented in [7]. The authors designed and carried out experiments on human subjects in which they attempted to assess the separate contributions of the prosodic, segmental, and voicing characteristics of parametric speech synthesis. They concluded that "although there have been large improvements in segmental intelligibility of synthetic speech in the past decade, improving the naturalness of speech generated by rule remains a significant challenge."

4.8 Future Challenges

Three elements of synthetic speech need improvement: prosody, prosody, and prosody. Language is rhythmic and musical. When the rhythm and tonality are unnatural, everything suffers, from intelligibility to naturalness and pleasantness. Some languages, including English, are intonational. When intonation goes awry, the listener is distracted and surprised, which reduces comprehensibility and the perceived quality of the speech.

One wishes that all textual material could be rendered orally, both for purposes of telephone communication (for example, calling your computer and having it read your e-mail) and as an aid to the visually impaired. The first of these may eventually be addressed as personal computers are developed that handle phone communication, or as videophones become more widely used.

Text-to-speech clever enough to read scientific formulae, computer programs, etc. is still in its early stages. A method for orally communicating the display on a computer monitor is likewise incipient. Text-to-speech intelligent enough to read poetry with the correct meter, and to read a script, play, or other highly formatted texts including electronic mail, are subjects for research that will stretch into the new millennium.

References

[1] Chomsky, N., and M. Halle, *The Sound Pattern of English*, New York: Harper & Row, Publishers, 1968.

[2] Vitale, T., "An Algorithm for High Accuracy Name Pronunciation by Parametric Speech Synthesizer," *Computational Linguistics*, V. 17, No. 3, September, 1991.

[3] Spiegel, M. F., "Coping with Telephone Directories That Were Never Intended for Synthesis Applications," *Proceedings of ESCA Workshop: Applications of Speech Technology*, Lautrach, Germany, 1993.

[4] McKeown, K. R., *Text Generation*, New York, NY: Cambridge University Press, 1985.

[5] Pisoni, D. B., "Perception of synthetic speech," In *Progress in Speech Synthesis*, Jan P. H. Van Santen et al. (eds.), New York, NY: Springer-Verlag New York, Inc., 1997, pp. 541–560.

[6] Lauretta, D., R. Rodman, and J. Antin, "The Effects of Familiarization on the Comprehension of Synthetic Speech in Telephone Communication," *Proceedings of the 34th Annual Meeting of the Human Factors Society*, 1990, pp. 189–193.

[7]   Nusbaum, H. C., A. L. Francis, and A. S. Henly, "Measuring the Naturalness of Synthetic
      Speech," *International Journal of Speech Technology,* Vol. 1, No.1, 1995, pp. 7–20.

5

Speaker Recognition, Language Identification, and Lip Synchronization

5.1 Speaker Recognition

> You mentioned your name as if I should recognize it, but beyond the obvious facts that you are a bachelor, a solicitor, a Freemason, and an asthmatic, I know nothing whatever about you.
>
> Sir Arthur Conan Doyle, "The Norwood Builder," *The Memoirs of Sherlock Holmes*

What personal characteristics of a speaker can a computer deduce based solely on that person's speech signal, without the necessity of speech recognition? This is a fundamental question in speech science. It is a pursuit worthy of the great Sherlock Holmes. The clues in speech are elusive and fleeting. Remarkable powers of observation and reasoning are needed to extract knowledge about the speaker. Nonetheless, speech can provide a wealth of information about a speaker.

From the speech signal alone good guesses can be made as to whether the speaker is male or female, adult or child. Indications of a person's mood, emotional state and attitude may be found in speech. Anger, fear, belligerence, sadness, indignation, reluctance, and elation may all be detectable in the speech signal.

Evidence of a person's nationality, region of upbringing, social standing, and level of education are present in speech. Traces of whether that person is speaking formally or casually, to intimates or to strangers, to persons of higher social rank or lower social rank, to children or to adults, to foreigners or to nationals, may appear in the speech signal.

An accurate portrayal of how the lips, tongue, mouth, and jaw of the speaker appear during the utterance of the speech may be inferred from the speech signal. The language a person is speaking may also be deduced.

If speech signal samples from a group of people are collected, compressed, and stored in a computer, it is possible for the computer to recognize members of the group by their voices at a later time. It is also possible to detect **impostors**—persons falsely claiming to be a member of the group. If an unknown speaker claims to be a certain person, the claim can be verified based on a comparison of their respective speech signals. If we have two samples of speech, it can be determined whether they were made by the same person. All of the operations mentioned in this paragraph are instances of **speaker recognition.**

Voice quality is one of several **biometrics** that can be used to determine who a person is. A biometric is a measurement of individuality based on a biological characteristic. Familiar biometrics are fingerprints, retinal patterns, handwriting patterns, facial features, blood types, and DNA.

5.1.1 Speaker Recognition Versus Speech Recognition

Speaker recognition is the complement of speech recognition. In speech recognition, the computer tries to extract linguistic information from the speech signal to the exclusion of personal information. That is especially important for speaker-independent speech recognition. Speaker recognition requires the computer to focus on the speech characteristics unique to the individual, disregarding the actual words spoken.

People recognize people by their voices. We expect to be recognized when we phone home. A phone call beginning "Hello, Doris, this is your husband Jack," would seem odd unless Jack had been out of touch for a long time. More normal in such cases is "Hi, it's me."

We not only recognize the voices of intimates, we are also able to recognize the voices of celebrities such as news anchors, sports heroes, politicians, and movie stars. We even recognize their identity when we can't hear clearly enough to know what they are saying. It is plausible to conclude that the speech signal contains sufficient idiosyncratic information about the speaker to allow correct identification.

It shouldn't be at all surprising that people have unique voices. Voice quality is dependent on a number of anatomical features. Among these are the dimensions of the oral tract, pharynx, and nasal cavities; the size and shape of the tongue and lips; the position of the teeth; the tissue densities and elasticity of these organs; the size of the larynx; and the tension and density of the vocal cords. If we could measure these physical features accurately we would undoubtedly discover that no two individuals are alike. Vocal tract anatomy is as accurate a distinguisher of individuals as fingerprints or DNA.

Vocal tract anatomies are, however, much more difficult to compare than fingerprints or DNA patterns. Comparisons could be made using the advanced and costly scanning technologies available to the medical industry, but that's impractical in most instances. Rather, we fall back on indirect evidence of these vocal tract properties, which may be revealed by analyzing the speech signal.

In practice there are similarities between speech recognition and speaker recognition. Both have a **training phase** in which speakers' voices are registered or enrolled, and an **operational phase,** when actual recognition occurs. With speech recognition, some kind of a model is made of whatever linguistic units the speech recognizer uses, as we saw in Chapter 3. With speaker recognition, a model is made of pertinent, speaker-defining voice quality characteristics.

In the operational phase, the current utterance is compared with the pre-stored models to determine either which linguistic unit was uttered, in the case of speech recognition, or which speaker actually spoke, in the case of speaker recognition.

5.1.2   Types of Speaker Recognition

Law enforcement and military security authorities were among the first to make use of speaker recognition technology. Here are three typical scenarios:

- *Scenario 1:* A death threat is received on X's answering machine, and shortly thereafter X is murdered. The phone call came from a building to which only six people had access. Each of the six is interviewed by the police. The interviews are recorded, and each suspect's voice is compared with the voice on the answering machine. The suspect whose voice is the closest match is detained.

- *Scenario 2:* Same as scenario 1, except for the possibility that someone other than the six who had building access may have placed the call. Does the closest matching voice from among the six belong to the murderer, or could it be someone else?

- *Scenario 3:* A wiretap recording is obtained of an unknown person leaking military secrets to a foreign agent. There is also a wiretap recording of a known suspect. Were the voices on the wiretaps from the same person?

The first two scenarios are cases of **speaker identification.** In such cases, there is a pool of voices against which the unknown voice is matched. If it is known *a priori* that the unknown is a pool member, and it is only a question of which pool member, we have **closed set speaker identification.** This is like Scenario 1.

The second scenario is more difficult. After a closed set identification is made, it must be further verified that the pool member chosen matched the unknown closely enough to exclude the possibility of an outsider. This is termed **open set speaker identification.**

The third scenario is a case of **speaker verification.** There is no pool of speakers on which to base the comparison. It is a question of: "Were two voice recordings made by the same person?" Speaker verification is used extensively in security applications to control access to a facility or data base, or to verify credit card or cellular phone usage. In such cases users volunteer to make the initial recordings of the training phase, a process called **voice registration.** The operational phase is entered whenever a user requests access, at which time they make a new recording for comparison. Speaker verification is ascertaining that a person is who they claim to be through a process of matching a current voice recording with one(s) made on a previous occasion.

As indicated in Scenario 3, the user doesn't necessarily volunteer to record his voice for the "training phase." In this case a known person is surreptitiously recorded, which serves as the voice registration. The operational phase is "verifying" another recorded voice against the registered one.

These definitions are summed up in Table 5.1.

5.1.3 Text-Dependent, Text-Independent, and Text-Prompted Speaker Recognition

If speaker recognition is based on voice recordings of the same utterance in both the training and operational phases, it is **text-dependent.** If the training

Table 5.1
Types of Speaker Recognition

Type Of Speaker Recognition	Description Of Process
Speaker identification, closed set	Determines which member of a set of speakers a voice belongs to
Speaker identification, open set	Determines which member of a set of speakers a voice *might* belong to, if it belongs to any member of the set at all
Speaker verification	Determines if a speaker is who he or she claims to be; equivalently, determines if two voice recordings were made by the same person.

utterances may be different than the operational utterances, the speaker recognition process is **text-independent.**

Text-dependent speaker recognition tends to be easier to implement and more reliable than text-independent recognition. The linguistic differences between training and operational utterances are less pronounced, though by no means absent, so the system can more easily focus on the idiosyncratic differences unique to each speaker.

Text-dependent systems require fewer training utterances. Several repetitions of the prescribed utterance are often sufficient. These systems are used most often by cooperative speakers in the kinds of security applications where passwords are ordinarily found, such as accessing computer systems. They are also used to control access to a physical facility such as a corporate site.

Text-independent systems allow any utterances to be used both in training and operation. They require many more seconds of training data to ensure that a wide variety of sounds are spoken, which in turn provides more information for constructing a model of the speaker's voice. Text-independent systems are suitable for applications where speakers may be uncooperative and not willing to repeat, say, a recorded bomb threat.

In such cases, a suspect may be recorded saying anything at all. This becomes the training data. The bomb threat is the operational utterance on which the speaker recognition is based. Thus the training phase need not precede the operational phase in time, nor does the speaker necessarily agree to, or cooperate with, the training process.

The most prolific application for speaker recognition is access control. It is generally text-dependent. The user registers by speaking a specific, identifying utterance (analogous to a password). To gain access, the same utterance is repeated. Such a system is easily foiled by a determined criminal who manages to record the person speaking the crucial utterance, and can then gain access by playing the recording to the system.

The use of several identifying utterances, one of which the system chooses at the time of entry, is only slightly more secure. It would force illegal entrants to obtain digital recordings of all the crucial utterances. The recordings would have to be digital to provide random access. At the time of entry, the recording of the particular utterance could be selected at a moment's notice and played back to the system.

To foil resourceful impostors a new kind of speaker recognition was devised called **text-prompted speaker recognition.** It is actually a hybrid of text-independent recognition and speech recognition. In such cases, the system prompts the user with a new key utterance on each occasion of attempted access. The system must be satisfied that both that the key utterance was repeated (speech recognition) and that it was spoken by an authorized person (speaker recognition). The system has the further advantage of being able to ask for more and more data until it is able to reach a conclusion with a high degree of certainty. The key utterances cannot be known in advance by a potential impostor, nor would it help the criminal to record random utterances of an authorized person, as it would if text-independent recognition were performed without speech recognition.

5.1.4 "Voiceprints"

Speaker identification using spectrographic evidence has been used in a number of criminal cases. My best guess at the moment . . . is that an expert may be wrong about one time in twenty in his positive identification of an unknown voice. . .

Peter Ladefoged, 1993

Speaker recognition began with **voiceprints** in the 1960s, a term whose coinage was based on the familiar *fingerprint*. In fact voiceprints were nothing other than speech spectrograms (Chapter 1). The registration phase consisted of having speakers utter ten commonly used words, such as *the, you, I,* etc.—presumably analogous to the ten fingers. The operation phase was conducted by human experts, who visually examined spectrograms of the same word uttered on different occasions, and determined a degree of similarity. Recognition was based on the aggregate similarities of the ten words. This was text-dependent recognition.

The experts looked at such features of the spectrogram as formant frequencies, formant bandwidths and trajectories, speech rates, degrees of nasality and so on. This superficial approach was no more successful in accomplishing reliable speaker recognition than it was for speech recognition. Much of the information contained in the speech signal must be obtained from precise measurements in the frequency domain, which is blurred even in wide band spectrograms.

Three other serious problems prevented voiceprints from ever being viable. The first problem was that intraspeaker variation often exceeded interspeaker variation for the parameters being considered. This was especially problematic because of the choice of short, common words as the text. The second problem was interexpert variation. Voiceprint reading can be subjective, and the experts themselves often disagreed as to the degree of similarity of two spectrograms. The third problem was the vulnerability of the system to trained mimics, who could duplicate another person's voice well enough to fool the spectrogram-based system.

Undaunted, or perhaps having no other recourse, the FBI continued to use spectrogram recognition well into the 1980s. They established procedures designed to reduce the interexpert variation by removing some of the subjectivity of the analysis, and they insisted on high-quality recordings as the basis for the spectrograms. Their system was also text-dependent with a 20-word (fingers and toes?) vocabulary.

One study of FBI cases found that of 2,000 voice comparisons, 1,304 produced low, or no confidence scores from examiners. Of the remaining 696 cases there were only three errors [1]. This remarkably low error rate, to the credit of the bureau, is due to the stringent procedures that prevented errors once a decision was made. The downside, of course, is that two-thirds of the cases led nowhere.

Although the use of spectrograms as voiceprints was in some sense a failure, it had the positive effect of forcing speech scientists and engineers to define the area of speaker recognition more carefully and to establish general guidelines for progress in the area. Several of these are summarized as follows[1]:

1. *Automate the operational phase.* By the mid 1980s sufficient computer power existed to make this feasible.

2. *Find techniques that reduce intravoice variation to an acceptable level.* This implies discovering how to characterize individual speakers more uniquely. This will also help with speakers who disguise their voice to escape detection.

3. *Find ways of reducing the dependency on a fixed vocabulary.* In other words, move toward text-independent recognition.

4. *Devise means for handling noisy signals.* We'll discuss this later on in Section 5.1.5.

5. *Devise means for coping with unknown or inadequate channels.* This will also be discussed in Section 5.1.5.1.

6. *Standardize testing of systems so the results of changes in procedures or technology could be measured.* We'll discuss this below when we get to Section 5.1.6 on performance evaluation.

5.1.5 Methods of Speaker Recognition

Speaker recognition, like speech recognition, is a two-pronged process. Measurable parameters of recognition must be identified, and a way of comparing parameter measurements is needed.

---

1. This information was extracted from FBI internal procedure papers, loaned to me by Mike Robertson of the North Carolina State Bureau of Investigation.

To recognize a face we take into account the shape of the jaw, the fullness and breadth of the lips, the size and shape of the nose, the height of the cheek-bones, the distance between eyes, the color of the eyes, etc. These are several of the parameters of facial recognition, used both by humans (subconsciously, for the most part) and computers. When we compare two faces we are actually comparing the aggregate of these parameters. This may be seen in remarks made around the crib like: "Oh, isn't it wonderful; she has her mother's blue eyes and her daddy's strong jaw."

Speaker recognition is no different. We must first identify parameters of the speech signal, paying heed to ones likely to reflect physical differences in speakers; and we must devise means for measuring sameness and difference among these parameters.

In a speaker recognition system the computer creates and stores models of known speakers' voices that are based on parametric measurements. When presented with speech from an unknown source, the system creates a parametric model in a similar vein, and then compares models. In the case of speaker verification, two models are compared—that of the unknown and that of the person who the unknown claims to be. If the models are sufficiently similar, the verification is made; otherwise the claimant is rejected.

For closed set identification, the known speaker whose model most closely resembles the unknown speaker's model is declared to be the unknown speaker, providing there are no other known models that resemble the unknown to a similar degree. Open set identification is like closed set identification, except that the unknown model that wins must also resemble the known model to a sufficiently high degree. In effect, open set identification is closed set identification plus an additional verification. If the verification fails, the unknown speaker is declared not to be any of the known speakers.

5.1.5.1 Parameters

Once the speech signal is sampled and quantized, it is ready for parameter extraction.

Pitch

The simplest parameter is based on pitch. Pitch extraction may be accomplished in several ways. For example, it may be determined by counting the glottal pulses (individual vocal cord vibrations) in one second in a spectrogram, as illustrated in Chapter 2. It may also be computed from the autocorrelation function. Such a function compares pieces of the signal in the time domain with one another. When values tend to be the same at different times—when they correlate—the time difference is a multiple of the pitch period, and the shortest such time that produces a high correlation is the actual pitch period.

The speech signal requires some preconditioning for the autocorrelation method to work accurately. This method, and several others for measuring the pitch period are discussed in [2].

The average pitch over an utterance is useful at a crude level of speaker identification. It can be used more or less successfully to distinguish among male, female and juvenile speakers, but it would not be successful in distinguishing among persons whose voices are at similar pitch levels. Also, pitch may be affected by the speaker's mood, and it can be deliberately modified by a speaker with ulterior motives.

Frequency

Speech may be represented in the frequency domain. A speech spectrogram is one such representation, and is generally computed from a discrete Fourier transform. Numerically, it consists of sequences of triples of numbers: time, frequency, and amplitude. Collectively, these numbers contain information about the individual speaker such as formant frequencies. These are related to the shape of the individual's resonant cavities, one of many distinguishing anatomical features of the vocal tract.

As noted above, early investigators attempted to use spectrograms pictorially to determine speaker similarity, often homing in on coarse features such as formant widths and shapes. Computers are able to perform more precise calculations on the raw numbers that underlie the spectrogram, and therefore produce more accurate results.

Linear Predictive Coefficients

Linear predictive coefficients are used throughout speech processing, including speech synthesis, speech recognition, and speech compression. It comes as no surprise to find them used in speaker recognition.

We saw how linear predictor coefficients are computed in Chapter 2. Combinations of these coefficients may be used directly as speaker-dependent features to build speaker models. It is also common to use transformations of the coefficients. Cepstral coefficients, reflection coefficients, and log area coefficients—all derivable from the LPC coefficients—have been used at one time or another to model the speech of individual speakers. (Details of their computation may be found in [3].)

For noisy environments delta-cepstrum coefficients are calculated by computing the differences between cepstral coefficients in each time frame. Any constant bias caused by noise or channel distortion would be removed. Relative spectral-based coefficients (RASTA) use a series of transformations to remove signal distortion. Stationary or slow-moving variations in the frequency domain, typical of noise, are detected and removed. Fast-moving variations, which are the result of the speech itself, are captured in the resulting parameters.

Other parameters too numerous or too arcane to mention have also been used at various times by various investigators. However, it would be safe to say that most speaker recognition systems in operation as of this writing use some kind of LPC-derived coefficients as the parameters for building speaker models.

5.1.5.2 Pattern Recognition

Once parameters are collected they must be integrated into a model that purports to represent the speaker. The models must be constructed so as to allow the computation of a distance measure between any two models that will reflect the degree of similarity between the speakers to whom the models belong. Several modeling techniques are discussed in this section.

Long-Term Averaging

In this technique, often used with pitch and/or cepstral coefficient parameters, a large number of feature vectors is collected for each known speaker. The average and variance of each component of the feature vector are computed, and a vector of mean (average) values, and a vector of the variances, are used to model each speaker. A similar model is made for the unknown speaker.

Many techniques exist for measuring the distance between vectors. The most common is the so-called Euclidean distance measure. It considers each vector to represent a point in a multidimensional space and computes the distance between the points. In a weighted Euclidean distance measure, more significance is given to dimensions that have low variance, since low variance indicates consistency and consistent measurements are often more reliable.

Long-term averaging is most useful for text-independent recognition, where large amounts of data are necessarily used to construct speaker models. The method would be ineffective if utterances were short because there wouldn't be enough data to be statistically significant.

The major weakness of long-term averaging is that each speaker's model consists of a single cluster of data represented by an average and variance vector. If the data contain multiple clusters of vectors, the variance will be very high. Since human speech is composed primarily of vowels, it is natural to expect feature vectors to form clusters, each one based on the pronunciation of a specific vowel. One way to get around this shortcoming is VQ.

Vector Quantization

A description of VQ, a form of vector "averaging," was given in Chapter 2. It is superior to long-term averaging because each speaker's model consists of several clusters of data, along with their centroids. VQ reduces these sets of vectors to a codebook, which provides an efficient way of building and comparing models of speakers.

VQ is used in several ways in speaker recognition. In some systems it is used simply to compress data. In other systems, VQ is a preprocessing step for other methods such as HMMs. However, the most common usage of VQ in speaker recognition is as a pattern-recognition scheme in and of itself.

For text-dependent identification and verification several codebooks are created or "trained" for each speaker, who speaks a prescribed text several times. This sequence of codebooks is the speaker's *template.* During the operational phase, the same prescribed text is spoken by the unknown person. The comparison is based on the accumulated differences between the unknown person's template, and each trained template, after a time-alignment process is used to remove variations in the speaking rate. In the case of identification, the person whose training template is most similar to the unknown template is identified as the unknown. For verification, the difference between the unknown and the template of the person the unknown claims to be, must be smaller than a predetermined criterion. The use of VQ for text-dependent speaker recognition is similar to the speech recognition of discrete words using dynamic time warping. (See Section 3.8.1.)

For text-independent speaker recognition a single codebook is associated with each speaker. However, that codebook is an accurate model of the speaker because it is formed from a much larger amount of speech than in the text-dependent case. For identification a comparison of the unknown's vectors is made directly with each codebook of each known speaker. The known speaker's codebook with the smallest accumulated difference is deemed to be the unknown. In the case of verification, the difference between the claimant's vectors and the codebook of who he claims to be, must not exceed a criterion value.

The VQ method introduces a new factor affecting performance: codebook size. With a codebook of one vector, VQ is the same as long-term averaging. Larger codebooks do a better job of characterizing a speaker's voice, but there is increased computational expense, and the danger of not producing results in real time, which is a significant factor for verification. (A 30-second wait to access one's bank account would be onerous.) On the plus side, VQ is valuable because it reduces the amount of data in the speaker's model drastically without much loss in accuracy.

Hidden Markov Models

HMMs are useful for modeling the stationary and transient properties of the speech signal. (They are described in Chapter 3, where they are applied to speech recognition.) They are appropriate for speech because speech sounds are both sustained, as with vowels, and ephemeral, as with many consonants. HMMs are able to represent signals that exhibit diverse behavior because of their probabilistic nature.

HMMs may be used for text-dependent speaker recognition. The process is similar to the way in which HMMs are used for small-vocabulary, discrete word speech recognition (see Figure 3.11). A single HMM is trained for each individual uttering the prescribed text. When an unknown person speaks the same speech, an HMM is created on the spot and compared with all the others. Commonly, the feature vectors used with HMMs are averaged through vector quantization and expressed as codebook values. Under favorable circumstances of low noise and clear channels, such systems may achieve the accuracy necessary for security applications.

HMMs have also been used for text-independent recognition. The "states" are trained to represent each person's pronunciation of the different phonetic classes, such as rounded vowels or nasal consonants. The HMM models for speaker independent recognition tend to be more highly interconnected because the training data are long and many sound types are repeated, leading to many more transitions than in left-to-right models such as Figure 3.11, which is designed to distinguish simply among different triphones. When all states have transitions of nonzero probability to every other state, the model is said to be **ergotic,** as illustrated in Figure 5.1. In essence, an ergotic model is insensitive to the temporal ordering of events.

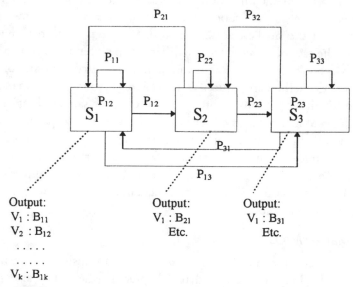

Figure 5.1 A three-state ergotic HMM.

During training, the parameters of the HMM are adjusted to best represent the salient features of each person's speech. One HMM is trained in this way for each known speaker. During the operational phase, it is determined mathematically which model is most consistent with the unknown input speech, and that model determines the unknown speaker, or confirms or denies a verification.

Neural networks

Neural networks are described in Chapter 3. They are computational models that attempt to emulate the human brain through interconnected nodes that behave like simple nerve cells. They are versatile devices and have been used for many different tasks including classification, which is how we saw them used with regard to speech recognition. They may be used in a similar way for speaker recognition.

In a typical system a neural network is created for each speaker and trained to be active (i.e., to give an output near 1.0) on input belonging to the speaker, and inactive (an output near zero) on all other speakers.) Clearly, each such network needs to be heavily trained both on its speaker and on a panoply of other speakers. (One of the drawbacks to neural networks is that they require extensive amounts of training.) The input to such networks may be any of the parameters discussed earlier. A typical input might be a vector of 10 LPC-derived cepstral coefficients.

To achieve speaker identification the set of input vectors from the unknown speaker is fed into the networks of all the known speakers. The network with the highest accumulated score identifies the unknown person. For speaker verification, the claimant's input vectors are fed through the network of the person who he claims to be. Verification occurs if the network returns a score that exceeds a preset threshold.

Another strategy using neural nets is to have one large network with one output per known speaker. During training, examples from each speaker are presented to the network. When a particular speaker's feature vectors are input, the network is tuned to return a value of one as the output corresponding to that speaker, and zero for all other outputs. During the identification stage the unknown set of input vectors is fed to the network and the output scores corresponding to each speaker are observed. The highest of those scores determines the identification. A verification operation is performed in the same way, except the winner's output value must exceed a threshold.

The use of a single large network works well for small populations of speakers, but it has two disadvantages. One is that if the size of the population exceeds a few dozen, the training times go way up, and the performance goes way down. As well, when new speakers are added to the population, the entire network must be retrained.

A variation of this method is to have a number of small binary networks—ones that make a distinction between only two speakers. The training time for each network is very short, although there are numerous networks, $N(N-1)/2$ to be exact, for a population of N speakers. Still, since each network is responsible for a small portion of the overall classification, each one could be highly specialized and could be expected to perform very accurately. To perform a speaker identification the unknown feature vectors are fed into every binary network. The output of each network is tallied and the speaker of greatest total score across all network pairs is considered to be the unknown.

Let's look at an example. In a population of three speakers, A, B, and C, we would have three binary networks: one trained to distinguish between (A,B), one between (A,C), and one between (B,C). If an unknown, X, is to be determined from among A, B and C the procedure would be as follows. Feed X into the (A,B) network and record the output score for A and the output score for B. Repeat the process with the (A,C) network and the (B,C) network. For example, suppose X was actually A. Then A would have a high score when X is fed into (A,B) and (A,C). In those cases both B and C would have low scores. When fed into (B,C), both B and C would have low scores. In total, A would be the winner.

The use of numerous binary detectors is suitable for closed set speaker identification with large populations. They may also be adapted to verification in the following way. Using the above example as an illustration, suppose we wanted to verify speaker A. A's input vectors would be fed to all binary networks in which A participates, namely (A,B) and (A,C). To be verified, A must score sufficiently highly over *all* other speakers in the population.

Neural net experts are taking rapid strides forward as I write. Progress is promising and it is likely that by the time you read these words, several variations of neural nets will have been invented, and dozens more applications using "traditional" neural net technology will have been put in place.

Segregating Systems

Segregating systems may be used for text-independent speaker recognition. The basic idea is to compare a specific sound class produced by the unknown speaker with the same sound class produced by the known speakers.

In the training phase, all input vectors are segregated into groups based on spectral characteristics associated with particular classes of speech sounds, for example, back, rounded vowels, i.e. the vowels in *root, rook. rote, and wrought.* For each known speaker, and each sound class, a VQ codebook, which serves as the speaker model, is created.

The operational phase is a two-step process. First, the input vectors of the unknown speaker are segregated and compressed into codebooks in the same way as in the training procedure. In the second stage the codebooks are compared. This achieves the effect of comparing a specific sound class produced by the unknown speaker with the same sound class produced by the known speakers.

Vectors may also be segregated using ergotic HMMs. Each state is associated with a phonetic category, thus achieving the phonetic-based segregation. A single such HMM is created during the training phase for each known speaker. In the operational phase the input vectors of the unknown speaker's utterance are tested against each HMM. Again, there is a two-stage process. In the first stage the likeliest sequence of states for that utterance is computed for each of the known HMMs. In the second stage the output vectors of the various likeliest state sequences are compared with the input vectors of the unknown speaker. The closest match determines the identification or verification required.

5.1.6 Noise

Noise is a recurring theme in computer speech technology. We have encountered and discussed the subject on previous occasions. (See Section 3.11.2.1.) The chief drawback of noisy environments is not the noise per se, but the variations in the noise. In speaker recognition, if both training and operational phases take place in the same noise, performance does not suffer unless the signal-to-noise ratio is low. If the noise is different between phases, or present in one phase and absent in another, then performance suffers drastically.

There are infinitely many potential sources of noise in speaker recognition. The noise may be classified as channel noise, ambient noise, or processing noise. Processing noise results from quantization error, as discussed in Chapter 2. (For the most part, "sampling noise" or aliasing can always be avoided by filtering the signal in accordance with the sampling rate—also discussed in Chapter 2.) Ambient noise may come from virtually anywhere in the testing and operational environment, including the speaker who may create noise vocally or by "stirring about" during the testing and/or operational phases.

The source of channel noise may be the microphone or transmission medium. Typical channel noise is additive, often heard as "static" or "hum" on the line. It may have a variety of sources. The other kind of channel "noise"— more accurately portrayed as channel variation—is the result of band-pass filtering, as with telephone channels, which typically exclude frequency components above 3,300 Hz and below 100 Hz. For example a speaker recognition system that trains speakers directly but attempts to verify them over phone lines may not work well. Moreover training on a landline but operating over cellular channels may also prove difficult because the transition media have different characteristics.

Speaker recognition systems may defend themselves against additive noise by preprocessing signals to separate the noise components of the signal from the speech component. To obviate channel variation, parametric features may be chosen that are immune, or only slightly affected, by the bandpass filtering typical of communication channels such as the public telephone network. As noted above, delta-cepstrum and RASTA coefficients are particularly effective in dealing with noise or channel induced signal distortion.

Finally, it has been suggested that the channel be modeled statistically on known signals. The model can be used to compensate for any future signal distortion. By choosing features that are resistant to channel effects or by removing channel effects before features are extracted, one can design systems for speaker recognition that will perform satisfactorily in less-than-perfect environments.

5.1.6.1   Channel Mismatch

If the training phase of speaker recognition occurs via one channel, the operational phase occurs through another channel, and the two channels have different bandpass filtering characteristics, lower recognition performance is likely to result. This situation typically occurs when the operational exemplars are recorded over the telephone, as in a bomb threat, and the training exemplars are obtained at police headquarters, as described above.

It is generally impossible to know the route through the phone system of any given telephone call, and differing routes may have differing filtering effects. Thus, the operational utterances are through an unknown channel, whereas the training utterances are through a known channel. To compensate for whatever effect the mismatch produces a technique known as **average filter compensation** may be used [3].

By assuming that the unknown channel's filtering effect is stationary—a reasonable but not always true assumption—a compensation factor for the channel mismatch may be calculated as follows. First, average spectra for the utterance over the *unknown* channel are computed (the bomb threat). Then average spectra of utterances over the *known* channel are computed (from police interviews). The averages may be taken in any of several different ways that needn't concern us. The result is a sequence of numbers representing the averages for both the unknown and known channel, say $A_{i,\,known}$ and $A_{i,unknown}$. A sequence of ratios of these numbers may then be computed.

$$R_{i,\,compensation} = \frac{A_{i,\,unknown}}{A_{i,\,known}}$$

The R_i are a measure of the channel filter mismatch. In practice the spectra sequence values of all the known utterances—the ones belonging to suspects—are multiplied by this factor to approximate the filtering effect that took place on the operational utterance (the bomb threat). At that point whatever speaker identification techniques are thought best may be applied. The reshaping of the spectra of known speakers obviates to some degree the distortion produced by the unknown speaker's channel.

5.1.7 Performance Evaluation of Speaker Recognition Systems

There are dozens of companies and many dozens of speaker recognition systems in the marketplace. The same cynical remarks about "manufacturers' testing" apply to speaker recognition as to speech recognition: *Don't believe a manufacturer's test results of its own products.* It's not that manufacturers are inherently dishonest. It's simply too easy to create conditions favorable for good performance in the laboratory that do not carry over to the outside, real world.

Product tests from neutral laboratories with no commercial axes to grind, such as those at NIST, are, of course, to be believed, providing you understand the test conditions. NIST is always scrupulously careful to made the test conditions clear. In the case of speaker recognition systems you will find that test results are always obtained from standardized databases of some sort, often available from the Linguistic Data Consortium. (See Section 3.11, footnote 3.)

5.1.7.1 A Particular Method for Evaluating Text-Independent Speaker Recognition Systems

In my laboratory at North Carolina State University, one of the several ways we evaluate the performance of text-independent speaker recognition systems is by using the Linguistic Data Consortium's TIMIT database. (Many other databases are available, all with catchy names such as SPIDRE, YOHO, and KING, that could be used in a similar manner.)

TIMIT contains speech from 420 speakers—230 male and 190 female—from eight dialect regions DR1–DR8. Each speaker was recorded speaking 10 sentences, two of which were the same for all speakers, called SA1 and SA2. The recordings are of high quality, having been sampled at 16 kilobits per second using a 16 bit quantizer.

To test a recognition system, we play the following game. We choose a dialect region and a gender, say the 22 DR1 male speakers. We pretend that all 22 are suspects in a crime and that the eight *different* sentences that each of them spoke were recorded by the authorities during an investigation. The eight sentences are used as the training utterances to construct a model for each speaker, as described previously. We further pretend that the SA1-SA2 sentences, the same for all speakers, were recorded as part of the criminal act, perhaps as part of a drug deal. These correspond to the operational utterances.

We begin with speaker 1 and use his SA1-SA2 utterances to construct a model of his voice, and we pretend he is an unknown suspect. We compare the model with the 22 models of "suspects," which of course includes his voice. If the system is good, the best match from among the suspects would be speaker 1 himself. In practice, however, speaker 1 may not be the best match. He might be second, third, or worse—and to that extent the recognition system is not good.

How can we quantify how good the match is? Just because the system determines speaker 1 as the best match doesn't mean the system is perfect. Would it do the same if it had to discriminate among 100 speakers, rather than 22? Moreover, what if it finds speaker 1 to be the second best guess? Isn't that a lot better than the tenth best guess? To get a figure of merit that addresses these questions we proceed as follows. We take the ranking of speaker 1—first through twenty-second—and divide it by 22, the total population.

That fraction is subtracted from 1.0 to give a figure of merit called **average percent eliminated (APE)**. For example, if speaker 1 was ranked first, the APE would be $1 - 1/22 = 0.954545$; if he was ranked second the APE would be $1 - 2/22 = 0.909091$, and so on. If speaker 1 is taken as the unknown voice, and speaker 1 is also deemed the best choice out of 22 suspects, then the 21 suspects who rank below speaker 1 can be thought of as no longer being suspects—especially if speaker 1 wins by a large margin. In other words, those 21 speakers, representing 95.4545% of the suspect pool, have been *eliminated* as suspects.

If the suspect pool were larger, say 100, and speaker 1 ranked first when he was the unknown speaker, then the APE would be 0.99. The better score represents the fact that speaker 1 ranked first under more difficult circumstances. Indeed, if speaker 1 was ranked second, yielding an APE of 0.98, one could argue that such a result is still *potentially* better than ranking first out of 22. Thus the population size is taken into account.

The experiment is repeated for speaker 2 through speaker 22. At the end, the 22 APE's are averaged for a more comprehensive figure of merit. During the experiment there are 22 model comparisons for each of the 22 speakers, for a total of 22^2 comparisons. This method is called "the n-squared test," which, together with APE scoring, is excellent for comparing different speaker recognizers since all variables are controlled, and the figure of merit is easily understood.

The entire test may be repeated for each dialect/gender group, as well as for the entire data base. Moreover, it can be used on noisy data, telephone channel data, cellular phone data, etc. to provide a basis for performance evaluation under specific usage conditions.

5.1.7.2 Evaluation of Speaker Verification Systems, the "Equal Error Rate," and the ROC Curve

For speaker verification systems a population of speakers creates training models that are used to allow or deny access based on a short operational phase utterance. Two types of errors are possible: a **false rejection** error occurs when a legitimate member of the population is mistakenly denied access; a **false acceptance** error occurs when an impostor is mistakenly allowed access. There are also two kinds of correct results: a **true acceptance,** where a legitimate person is allowed access; and **true rejection,** where an impostor is denied access. To evaluate a system one can observe and tabulate the percentages of each in a two-by-two array.

Percent of true acceptances	Percent of false acceptances (Error)
Percent of false rejections (error)	Percent of true rejections

Ideally, one wants high percentages in the upper left and lower right and low percentages in the upper right and lower left. A perfect verification system would yield the following results.

100	0
0	100

In practice, of course, one sees less than 100% and more than 0% in the respective boxes. The distribution of percentages depends on the acceptance criterion parameter. The more stringent the parameter, the higher the rate of false rejections, and the lower the rate of false acceptances, for example:

40	5
60	95

On the other hand a lax acceptance criterion might lead to these results:

90	20
10	80

Thus, one can control the trade-off of false rejection errors versus false acceptance errors by adjusting the acceptance criterion parameter.

In evaluating the performance of speaker verification systems, it is common to adjust (experimentally) the criterion parameter so that the false rejection error rate and the false acceptance error rate are equal—the **equal error rate** condition. In that case the respective corners of the box would be equal, for example:

70	30
30	70

The closer to 100 the upper left and lower right corners are, the more highly effective the verification system. This provides a figure of merit for rating speaker verification systems.

The result may be portrayed graphically as a *receiver operating characteristic* (ROC) curve [4]. This is a plot in which the x-axis is the probability of a false acceptance (letting in the bad guys), and the y-axis is the probability of true acceptance (letting in the good guys). For a given system, various settings of the acceptance criterion will, after testing, determine a set of points that define a curve from the lower left corner to the upper right corner of a square, as shown in Figure 5.2. The better the system, the more its curve will approach the upper left corner of the square, thus allowing an effective graphical comparison of the performance of different systems. (In a perfect system, the "curve" would follow the *y*-axis to the top horizontal line, then proceed along that line to the upper right corner.)

A low equal error rate may still not be satisfactory for some applications. Security-sensitive verification systems may be concerned with the lower limits of the false acceptance error rate in an effort to keep out impostors at any cost, including an inconveniently high rate of false rejections. (Better to annoy the good guys than let the bad guys in.) This would be achieved by a criterion parameter setting corresponding to point C in Figure 5.2. On the other hand, in an application in which impostors are not a threat, one would choose a criterion parameter corresponding to point R to avoid inconveniencing persons with valid access rights.

5.2 Co-channel Speaker Separation

When you were young—if you *are* young, enjoy this—you could converse with a person in a noisy room with ease. As we age such a feat becomes more diffi-

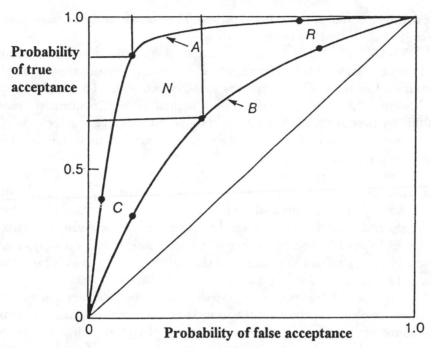

Figure 5.2 Two ROC curves. Curve *A* represents the performance of a speaker verification system superior to that represented by curve *B*. Points *C* are the result of a stringent acceptance criterion; points *R* are the result of a lax acceptance criterion; points *N* are the equal error rates of the two curves. The horizontal lines connecting the points *N* to the *y*-axis represent the respective false acceptance error rates at the equal error rate setting. The vertical lines from points *N* to the top horizontal line of the square represent the respective false rejection error rates at the equal error rate setting.

cult. Many older people eschew social contact in noisy environments simply because they cannot distinguish one person's speech from the noise in the background and are thus unable to have a meaningful conversation. In essence, they have lost the ability to perform *speaker separation* due to the hearing loss that accompanies aging.

Co-channel speaker separation is the extraction by computer of a particular person's speech from a milieu of sound comprising interfering speech and nonspeech signals The applications of speaker separation include hearing aids, reduction of cross-talk interference from neighboring communication channels, and the extraction of critical speech such as that of an airline pilot from the cockpit recorder of an airline in distress. The purpose of the extraction may be for speech recognition or speaker recognition.

Work in this area has been going on for just a little more than 20 years and can be considered still in its infancy. The simplest approach is based on filters. If the frequency of the interfering sound exceeds 4,000 Hz, the signal can be low-pass filtered at that level. Speech is intelligible at frequencies below 4,000 Hz. Even if the noise occurs in the frequency ranges of speech, if it is band-limited it may be filterable and still leave enough of the speech signal to make the speech comprehensible.

Unfortunately most situations involve two or more talkers, and filters will not work because both the desired signal and the interfering noise are in the same frequency ranges. This is where more sophisticated techniques are needed. Most practical work has undertaken the simpler task of separating two-talker co-channel speech where one talker is stronger. This would occur typically when the strong talker has control of a channel that is being assailed by cross channel interference, as might occur with a weak AM radio signal.

One approach is to process the speech signal in the frequency domain and attempt to estimate the pitch of the stronger talker. Once the pitch is known, the harmonics are also known—they are multiples of the pitch frequency. Knowing the harmonics means being able to locate formants, and that means knowing the vowels of the stronger talker and where they occur. Other harmonics are suppressed so that the end result is an enhancement of the stronger talker's speech, and in the best cases, total elimination of the interference.

The weaker talker's speech may be obtained by subtracting out or suppressing the stronger talker's signal from the total signal. This is useful when, for example, both a pilot and copilot are heard talking simultaneously on a recovered flight recorder.

The two speaker problem may be extended to multiple speakers by identifying the pitch ranges of each, and on that basis enhancing individual speech. Success is limited as the separation process becomes more tenuous with more speakers.

The goal of co-channel speaker separation, then, is to process by computer the various signals and recover each talker's speech. As well, the process seeks to eliminate nonspeech noise, especially if a talker's recovered speech is to be used as input to a speech or speaker recognition system.

5.3　Language Identification

All the Oriental nations jam tongue and words together in the throat, like the Hebrews and Syrians. All the Mediterranean peoples push their enunciation forward to the palate, like the Greeks and the Asians. All the Occidentals break their words on the teeth, like the Italians and Spaniards. . . .

Isidore of Seville, seventh century C.E.

Traveling, they say, broadens one's outlook. Foreign travel is also an opportunity to hear languages other than English spoken. For those who have an interest in human speech, a fun game to play while traveling is to eavesdrop on conversations at international crossroads such as the Eiffel Tower or the Vatican and try to figure out what language is being spoken.

If you know a foreign language, you inevitably recognize it when you hear it. Even if you don't know a foreign language well enough to understand or speak it, you may recognize it anyway, based on casual contact. Many non-Spanish speaking Americans from western states can identify Spanish when they hear it because they have been exposed to it so much.

The most fun for linguistic types such as myself is not to know a language per se and try to identify it based on its sound. This isn't as impossible as it seems. Many languages have characteristic phonetic patterns that help identify them. For example, Russian, and other Slavic languages such as Polish, have a "palatalized" sound—consonants followed by /y/, as in Russian *nyet* meaning "no." French sounds nasal because nasal vowels occur frequently in French. French also makes much more use of the voiced palatal fricative /zh/ (the medial consonant in *measure*), and that contributes to its character.

The question naturally arises as to whether computers can be programmed to identify an unknown language based on a short segment of speech. Automatic language identification by computer is a goal of the global community. Telephone companies would like to identify the language of foreign callers and automatically route their calls to operators or interpreters who know the language. A multilanguage machine translation program needs a language identification front-end that can determine the input language and choose the correct translator. Agencies that monitor the airwaves around the world have a need to identify the language they are monitoring. Emergency operators such as the ones that answer 911 calls in the United States often encounter foreign languages and while the emergency services generally provide access to interpreters of most of the common languages in their areas, it is not always easy to determine what the unknown language is. Horror stories surface from time to time about long delays caused by the operator playing prerecorded greetings in the wrong language and trying out several interpreters of yet other wrong languages before finding someone who recognizes the language of the distressed caller.

Before looking at how well computers can be programmed to identify languages, it is instructive to see how well people do. There is little research in this area. In one case several investigators from the Oregon Graduate Institute conducted two experiments on language identification by humans [5]. Speakers of English were exposed to the speech of nine languages[2] and then asked to identify speech from an unknown language chosen from among the nine. Average performance was at first 50% and with further exposure rose to 54%. Roughly, then, people get it right half the time on a nine language database.

5.3.1 Four Computational Approaches to Language Identification

Researchers have devised four methods of attacking the language identification problem. All of them are innovative; none of them yet have performed well enough for practical field applications.

5.3.1.1 The Speech Recognizer Approach

Like humans, computers are best at language identification when they "know" the language. When I say a computer "knows" a language, I mean that the computer is programmed to recognize speech in that language, as discussed in Chapter 3. If every language we wished to identify had a corresponding recognition system, we could treat computer *language* identification like computer *speaker* identification, where the individual language is analogous to a single speaker, and the speech recognizer for that language is analogous to the computer model of the speaker.

In closed set language identification, utterances from the unknown language are given to each of the speech recognizers for the known languages. The speech recognizer that recognizes the unknown speech with the highest confidence is that of the unknown language, providing there are no close runners-up. In open set language identification, the unknown speech must be recognized at a sufficiently high confidence level to verify that it is indeed the language being spoken and not a language unknown to all the recognizers. Figure 5.3 illustrates the process.

This method has proven as good or better than any other that has been tried. Unfortunately, its implementation depends on the availability of good speech recognition systems for all the candidate languages. The number of world's languages that have had speech recognizers designed and built for them is relatively small, numbering in the dozens at most. The task of building an effective continuous speech recognizer is prodigious and expensive. Less com-

---

2. The languages were Farsi (spoken in Iran), French, German, Japanese, Korean, Mandarin Chinese, Spanish, Tamil (spoken in India and Sri Lanka), and Vietnamese.

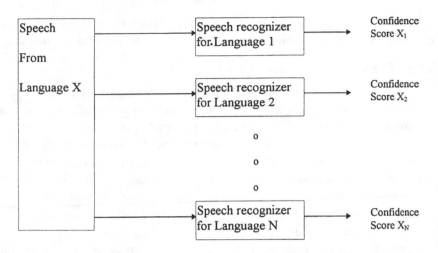

Language X = Language K, where K= the index of the maximum X_i

Figure 5.3 An illustration of the speech recognition method of language identification.

monly studied languages from poor countries, such as Albanian and thousands of others, have no speech recognition systems. Other methods are needed for computer identification of such languages.

5.3.1.2 The Language Model Approach

In the experiment with human listeners mentioned above, subjects were asked on what basis they decided what language they were hearing. Prominent among the answers they gave was that they listened for sound combinations characteristic of a particular language, such as /shp/ and /pf/, which occur word initially in German. The language model approach embodies this idea computationally.

A phone-based language model describes a language in terms of the statistical distribution of its allophones (sounds). For example such a model for English would reflect the fact that /st/ is a commonly occurring consonant cluster that may be followed by a vowel or /r/, but never /l/ (within the same syllable). To do language identification, models are built for all the candidate languages. The unknown language is analyzed into phonetic strings and the patterns of sounds are compared with each of the language models. The closest comparison is deemed to be the unknown language, at a certain confidence level.

This method has proved about as effective as the speech recognition approach. As with the latter, the principal drawback is the vast amount of resources required to build language models. Hundreds of thousands of utterances must be analyzed to create useful models. Only languages such as English or Japanese exist in an economic environment where such resources are available.

5.3.1.3 The Acoustic Model Approach

Computers can also be programmed to recognize languages by analyzing the acoustic features of the language, much as humans do when they deduce that an unknown language is German because it sounds guttural. (German has a velar fricative phoneme that leaves this impression.) Speech from the unknown language can be analyzed into a series of feature vectors based on the same kinds of acoustic properties used in speech and speaker recognition. This includes LPC derived coefficients, cepstral coefficients, formant locations, and pitch contours. The feature vectors of the unknown language are compared with those of various known languages.

Prosodic features such as pitch and intonation may provide important clues as to the language spoken. A tone language such as Mandarin Chinese or Thai is heard by the human ear as "singsong" because each syllable has its own pitch. Such languages are easily distinguished from nontone languages such as Japanese or English. On the other hand, Japanese is a language possessing the prosodic characteristic that every syllable is of equal duration. It gives Japanese a characteristic rhythmic sound that people recognize without knowing a word of Japanese. I once took a course in Japanese while preparing to teach in Kyoto for a semester. The instructor kept a baton and would mark time like an orchestra conductor as we students recited, forcing us to pronounce each syllable on the beat. Such prosodic clues can be exploited in computer language identification, especially in combination with other kinds of acoustic features.

VQ, HMMs, and neural nets have all been used by various investigators to compare properties of an unknown language with properties of known languages in attempting identification. Success as of this writing has been limited, with few researchers reporting accuracies above 80% from language pools of more than eight languages.

An advantage of this approach is that building an acoustically based model of a language is less computationally demanding than building a language model or speech recognition system. Moreover, if a language model is available for certain candidate languages, it can be used in conjunction with an acoustical model to yield improved accuracy. An excellent description of this method may be found in [6].

5.3.1.4 The Speaker Recognition Approach

This is a "good news, bad news" method, described in [7]. Speech samples are collected from each language from, say, 10 speakers. Suppose there are eight languages, so we have speech from 80 speakers. We now take the speech of an unknown language by an unknown speaker (probably not in the original set of speakers) and do text-independent *speaker* identification, ranking all 80 speakers according to their similarity to the unknown speaker. For each language, we add up the similarity scores for its speakers. The best score wins, subject to the usual condition that there are no near runners up. The good news is that performance using this method is comparable to other methods, as reported in [8]. Even better news is that the process may be accomplished using relatively inexpensive, accessible software.

The bad news, ironically enough, is that it works so well. Ideally, speaker recognition and language identification should be independent of each other. That is, we should be able to recognize a speaker irrespective of what language is spoken; and we should be able to identify a language irrespective of what speaker is speaking it. If we understood more completely the spectral properties of speech related specifically to the speaker, and those related specifically to the language, speaker and language identification wouldn't mix, and both would work far better than they do currently. The success of the method, then, is an indication that much research needs to be done in both speaker recognition and language identification.

A language identification system must not be confused by the diversity of speakers that produce utterances, or, within limits, by dialectal variation. An effective system ignores properties of the speech signal whose source is the individual speaker and instead focuses on features that vary across languages. The "shp" and "pf" consonant clusters that contribute to the character of spoken German can be heard in the speech of men, women, and children, and it is a matter of isolating the pertinent acoustic patterns and ignoring speaker specific parameters such as frequency ranges of the voiceless consonants.

Many of the same problems found in speech and speaker recognition also occur in language identification. If the signals are noisy, performance will decrease. Performance will also suffer if language models are built from direct speech but tested on telephone speech, or vice versa. Despite the difficulties some researchers have reported accuracies as high as 90% in closed set identification on 11 languages using realistic data from the Oregon Graduate Institute.[3]

5.3.2   Performance Evaluation of Language Identification Systems

Twenty years ago testing speech processing systems was as chaotic as law enforcement in the American west 120 years ago. Just as any gunman could put

---

3. I am referring to the Oregon Graduate Institute Telephone Corpus (OGI-TS), to be discussed in Section 5.3.2. Information is also available on the World Wide Web at http://www.cse.ogi.edu.

on a badge in Tombstone, Arizona, any company could develop a system—speech recognition, language identification, whatever—test it in their own lab on their own data and publish the results. Often the results were neither duplicable nor verifiable, so from a scientific point of view they were worthless, though this didn't prevent the bandying about of performance statistics in an attempt to gain market shares.

In the early 1980s, much to their credit, the then National Bureau of Standards, together with several forward-looking companies such as Texas Instruments, began to develop standardized databases and methodologies for testing speech processing systems. Nowadays it is *de rigeuer* to test a system using a standardized database.

The Oregon Graduate Institute collected telephone speech from 11 languages in 1992 for the purpose of evaluating language identification systems. The languages were English, Farsi, French, German, Hindi, Japanese, Korean, Mandarin, Spanish, Tamil, and Vietnamese. The corpus contains both short utterances such as numbers, days of the week, and names of languages; and fluent continuous speech elicited by asking subjects to respond to statements like "tell us something about your hometown." Utterances from about 2,000 speakers for a total of 23 hours of speech have been collected.

To assist researchers in the language identification field, the utterances were all transcribed orthographically. In addition some of the data were selected for phonetic transcription, carried out by professional linguists. This corpus is available to the general public.

There are many ways to use this corpus to evaluate a language identification system. Here are two of many possibilities. Suppose we train a collection of neural networks to distinguish between any two of the eleven languages. We would end up with 55 neural networks—the number of combinations of 11 objects taken two at a time—for example, one that is trained to distinguish Mandarin from Vietnamese, another trained on Mandarin and English, another trained on Vietnamese and Farsi, and so on until we have accounted for all the pairs. Each individual neural net is highly specialized for its two languages, having been trained on both positive and negative exemplars. That is, the neural net that distinguishes Mandarin from Vietnamese is given data from Mandarin and trained both to recognize that the exemplar *is* Mandarin and *is not* Vietnamese, and vice versa. Such neural nets perform well on their own two languages, distinguishing between them correctly over 90% of the time at high confidence levels. When given languages other than the two they were trained on and forced to choose the neural net performs at much lower confidence levels.

To test such a system (or to implement it) one feeds an unknown language to all 55 neural nets simultaneously and observes the output of each along with the confidence level. Suppose the unknown language is Mandarin.

Then ten of the neural nets, namely Mandarin/English, Mandarin/Farsi, etc., would show relatively high confidence levels for Mandarin, while the remaining 45 nets would show relatively low confidence levels for whatever language they chose. These results may be analyzed statistically and used to perform language identification. If another similar system comes along, it can be tested in an identical manner, so that the systems can be compared in a valid, measurable, repeatable way.

A second kind of system might follow the APE method described in Section 5.1.7.1 for speaker recognition. In this case the language identification system to be evaluated is exposed to various unknown languages and its guesses are ranked from one to eleven. For a given trial, if the unknown language was ranked one, the APE score would be $1 - 1/11 = 0.909$; if it were ranked second the APE would be $1 - 2/11 = .818$, and so on. After a large number of trials, the average of the APE scores would provide a figure of merit for the identifier. Moreover, the experiment could be duplicated for a different language identification system that operated in a similar manner, providing a valid means of comparison.

There are many different types of language identification systems in existence and several other language databases than the OGI corpus against which these systems could be tested. What is important is that some standardized database be used along with a well-defined testing procedure so that results may be replicated, this being the hallmark of the scientific method.

5.4 Lip Synchronization

Bodily functions are generally taken for granted until they fail. Then off we go to the doctor to fix that back that no longer bends, or those eyes that falter on the fine print. You might never give much thought to lip synchronization until one night on the late show you see a bunch of guys fleeing a giant wombat, and the words they're shouting are totally out of sync with their lip movements, an effect as eye-catching as the monster.

Lip synchronization is where a computer takes speech as input and outputs parameters for lip, jaw and tongue movements that correspond to the speech. The output is used to animate a face that appears to be speaking the speech in perfect sync. The principal applications are realistic cartoon animation and visual assistance to the hearing disabled.

Lip synchronization may be done with or without an accompanying text. When the text of the speech is known, the voice may be machine generated (i.e., synthetic) or come from a human actor. Synthetic speech is the easiest to lip sync because both the phonemes and the timing of their occurrence is known to the computer. Lip synching becomes a matter of using that information to animate a face correctly.[4]

Lip synching human speech with text is a more challenging problem The phonemes are readily determined—it's part of the text-to-speech process discussed in Chapter 4—but the timing of each phoneme's occurrence must be determined accurately for realistic lip sync. Timing information must be extracted from the speech signal using the kinds of techniques discussed throughout this book. For example in [9] a technique for extracting timing parameters using LPC is explained. Other researchers have used HMMs to achieve the same effect.

The most difficult lip synching is without text. In this case the computer must determine both the phonemes and their timing. In essence the computer does speech recognition while attempting to keep track in time of the beginning, middle, and end of the phonemes as they blend one into the other during continuously spoken speech. Most of the difficulties of recognizing continuous speech pointed out in Chapter 3 plague this approach to lip synchronization, which is well described in [10].

5.4.1 Visemes

Lip synching without text is similar to speech recognition but is both more difficult because timing must be considered, and easier because certain distinctions among phonemes need not be made. For example, lip synching need not distinguish between /p/, /b/, and /m/ because all three phonemes correspond to a lips-closed position. Similarly, one needn't distinguish between /f/ and /v/, /i/ and /e/, and so on, because the respective mouth positions are nearly the same. We say that /p/, /b/ and /m/ form a single **visual phoneme** called a **viseme.** Just as phonemes are the distinctive sound units of language (see Chapter 1), visemes are distinctive mouth shapes of language.

Unlike the phonemes, which are determined by the language system and are *discovered* (as opposed to *defined*) by linguists through linguistic analysis, visemes are defined by an animator or graphics expert to suit the purposes of a particular application. For purposes of lip reading, where it is important to distinguish among /i/ /ā/ and /e/ (the vowels of *bid, bade,* and *bed*), it may be suitable to define three slightly different visemes to correspond to these phonemes, and the total number of such visemes may number in the several dozens. For

---

4. The problem of actually animating a face to appear as realistic as possible is a challenging problem that is currently under investigation. For much more information, see the Internet site at http://mambo.ucsc.edu. Since this book is about voice technology, I will gloss over this very interesting area of computer graphics and focus on the speech processing side of the problem.

purposes of lip synching for animation, one viseme for /i/, /ā/ and /e/ suffices. Table 5.2 presents 13 cardinal visemes suitable for lip synching English, seven principally for vowels but including some consonants, and six for consonants.

Table 5.2
Visemes Suitable for Lip Synching English[5]

Viseme	Description	Examples
I	Lips spread wide but close together, as in a smile; jaw high	b*ee*t ch*ee*se
E	Lips spread and somewhat apart; jaw medium height	b*i*t bl*a*de b*e*t
AE	Lips spread and well apart; jaw low, some tongue visible	b*a*t
A	Lips somewhat spread and well apart; jaw low	b*a*lm b*u*t p*o*t
O	Lips somewhat rounded and apart; jaw medium height to low	s*aw* n*o*te
U	Lips rounded and close together; jaw high	p*u*t cr*u*de *w*in *r*eap
ə	Lips somewhat spread, somewhat apart and relaxed; jaw medium height	it*e*m
P	Lips together; jaw high	*P*aul *b*all *m*ock
F	Lower lip touching upper teeth; jaw high	*f*an *v*an
TH	Lips spread and somewhat apart, tip of tongue showing between teeth; jaw medium height	*th*igh *th*ough
T	Lips spread and somewhat apart; jaw medium height	*t*oll *d*ale *n*ot sin*k* zin*c* *y*ear *l*eap
SH	Lips extended and somewhat apart; jaw medium height	*sh*y vi*s*ion *ch*unk *j*unk
K	Lips spread and somewhat apart; jaw medium height to low	*k*ill *g*old ri*ng*

---

5. Dipthongs are not shown. They are dynamic and composed of at least two visemes, one indicating the facial shape at the start of the dipthong, the other the facial shape at the end of the dipthong. Also /h/ assumes the viseme shape of the vowel that follows it, so its viseme is different in *heat, hat, hut,* and *hoot,* for example.

(Of course animators are free to do "in betweening" and create other mouth positions than the basic 13.) The corresponding sounds are shown in boldface italics in example words. (You may refer to Tables 1.3 and 1.4 in Chapter 1 for the actual phonetic symbols.)

As noted, this particular set of visemes is somewhat arbitrary—finer or coarser distinctions are possible—but covers the principal mouth shapes fairly well. Four of the visemes are illustrated in Figure 5.4.

5.4.2   Mapping Directly From the Speech Signal to Mouth Shapes

Research scientists at North Carolina State University, including Drs. David McAllister, Donald Bitzer, and myself, have for the past several years been working on an innovative solution to the text-free lip synching problem [11]. Our approach is to analyze the speech signal spectra and extract parameters from them that correlate directly to mouth positions. This method eschews speech recognition, bypassing both the difficulties inherent in recognizing individual phonemes and determining when, in time, they occur.

The essence of our discovery is that the shape of the spectrum of the speech at each individual glottal pulse is correlated to the shape of the mouth—the viseme—at that time. The spectra are derived by taking the DFT of the time domain signal at 5-ms intervals using a window of 20 ms length. (See Chapter 2 to review these concepts.) These spectra are "smoothed" and otherwise processed to remove idiosyncratic information such as transient vibrations from the glottal pulse. Their shape is measured by taking *moments,* mathematical functions borrowed from statistical analysis where they are used as a simple method for describing the shape of the distribution of a probability density function.

In particular we treat the first moment (equivalent to the mean) and the second central moment (equivalent to the variance) as independent variables and the parameters that control the visemes as the dependent variables.

Figure 5.5 shows the location of six vowel visemes in a space where the x-axis is the first moment and the y-axis is the second central moment. The fig-

Figure 5.4   An illustration of the visemes E, I, A, and U (from left to right).

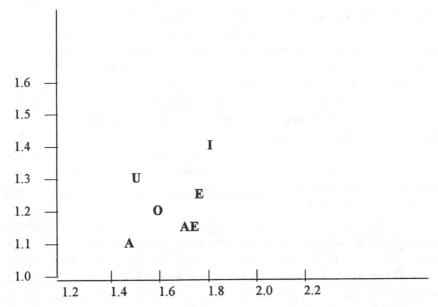

Figure 5.5 A plot of six vowel visemes in a space where the x-axis is the mean and the y-axis is the variance of the average spectra for the sounds comprising each viseme.

ure shows that the visemes are separated from one another; each location is associated with the mouth shape appropriate to the viseme.

Because our interest is in the speech processing aspect of lip synchronization, we have taken a simplistic approach to the graphical creation of the visemes. We control mouth shape through four parameters—the dependent variables—and it is those that determine the actual viseme. The parameters are: **jaw, flare, corners,** and **edges.** They are illustrated in Figure 5.6.

Jaw controls the height of the jaw, a reflection of tongue height. Thus the jaw is high for the viseme I and low for the viseme A. **Flare** is one of the

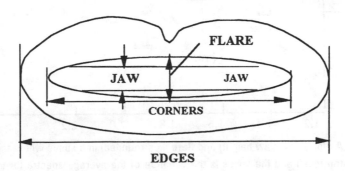

Figure 5.6 The mouth shape parameters **jaw, flare, corners,** and **edges.**

parameters that controls lips-spread versus lips-rounded. For the viseme I, **flare** would have a small value; for the viseme O it would have a larger value. **corners** is a measure of the point where the lips actually touch. For example, **corners** would be very small for the viseme P because the lips touch everywhere from the center out. It would be large for the visemes AE and A, where the mouth is nearly wide open. EDGES measures the overall width of the lips and is used as one of the parameters that controls lip rounding and extrusion. **Edges** would have a larger value for the visemes I and P than for U or SH.

The various parameters separate well in moment space, where again the *x*-axis is the first moment (mean) and the *y*-axis is the second central moment (variance) of the average vowel spectra. Figure 5.7 illustrates the locus of **jaw** and **flare** in such a space.

The data shown are averages for one speaker. The sounds of different speakers occur in different parts of the moment space, but are nonetheless separated, meaning that the process is currently speaker-dependent. Before determining the actual mouth shape, the particular correlations between location in moment space and mouth parameter values for each individual speaker must be known in advance. Thus, like speaker-dependent speech recognition, the lip sync system must be trained by each user to discover the locus in moment space

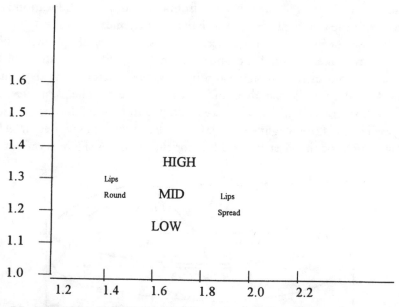

Figure 5.7 A plot of **jaw** (jaw height) and **flare** (lips rounded) in a space where the *x*-axis is the mean and the *y*-axis is the variance of the average spectra for the sounds comprising high mid and low jaw heights, and rounded and unrounded lip shapes.

of their viseme parameters. The training, fortunately, can be accomplished quickly, in a matter of minutes, and consists of having the user speak words containing the various sounds that comprise the various visemes.

5.5 Future Challenges

The most immediate challenge in speaker recognition is a speaker verification process that will be accurate enough so that financial transactions could take place under it aegis. That means very low impostor rates, perhaps as small as 0.001% (1 in 100,000). Moreover, that rate must be achieved with a relatively low rate of false rejections, ideally less than 0.1% (1 in 1000). These error rates must prevail in the face of the intraspeaker variation caused by aging, illness, and emotional state, as well as with use over various communication channels such as cellular phones.

The future challenges in language identification are perhaps less daunting than in speaker recognition because the field is relatively new. Three-quarters of the world's population speak but 20 languages. I would lay down the following challenge: a language identification system capable of handling telephone speech and distinguishing among the 20 most commonly spoken languages with an error rate under 1%, to be achieved by 2001.

The goal we have set for ourselves in lip synchronization is to have speaker-independent, language-independent lip synchronization—that is, any speaker of any language can pick up a microphone and begin talking, and the system will begin lip synching within two seconds. The lip synchronization must be accurate enough to permit hearing-impaired persons to lip-read the animated face with the same degree of fluency and comprehension, or even more, than with a human speaker's face. Our aim is to achieve this goal early in the 21st century.

References

[1] Koenig, Bruce E., "Spectrographic Voice Identification: A Forensic Survey," *Journal of the Acoustical Society of America*, Vol. 79, No. 6, 1986, pp. 2088–2090.

[2] Rabiner, L. R., and R. W. Schafer, *Digital Processing of Speech Signals*, Englewood Cliffs, New Jersey: Prentice-Hall, Inc., 1978.

[3] Klevans, R., and R. D. Rodman, *Voice Recognition*, Norwood, MA: Artech House, Inc., 1997.

[4] Wickens, C. D., *Engineering Psychology and Human Performance*, Glenview, IL: Scott, Foresman and Company, 1984.

[5] Muthusamy, Y. K., E. Barnard, and R. A. Cole, "Reviewing Automatic Language Identification," *IEEE Signal Processing Magazine*, October 1994, pp. 33–41.

[6] Hazen, T. J., and V. W. Zue, "Segment-based Automatic Language Identification," *Journal of the Acoustical Society of America,* Vol. 101, No. 4, 1997, pp. 2323–2331.

[7] Li, K. P. "Automatic Language Identification Using Syllabic Features," *Proceedings of the IEEE International Conference on Acoustics, Speech and Signal Processing,* Vol. 1, 1994, pp. I–297-I–300.

[8] Muthusamy, Y. K., N. Jain, and R. A. Cole, "Perceptual Benchmarks for Automatic Language Identification," *Proceedings of the IEEE International Conference on Acoustics, Speech and Signal Processing,* Vol. 1, 1994, pp. I–333-I–336.

[9] Lewis, J. P. "Automated Lip-Sync: Background and Techniques," *Journal of Visualization and Computer Animation,* Vol. 2, No. 4, 1991, pp. 118–122.

[10] Lavagetto, F. "Time-Delay Neural Networks for Estimating Lip Movements from Speech Analysis: A Useful Tool in Audio-Video Synchronization." *IEEE Transactions on Circuits and Systems for Video Technology,* Vol. 7, No. 5, 1997, pp. 786–800.

[11] McAllister, D. F., R. D. Rodman, D. L. Bitzer, and A. E. Freeman, "Lip Synchronization of Speech," *Proceedings of the European Speech Community Workshop on Audio-Visual Processing,* 1997. pp. 133–136.

6

Applications in Speech Recognition

> You have such strong words at command, that they make the smallest argument seem formidable.
>
> George Eliot, *Felix Holt, The Radical.*

What were they thinking, Arthur C. Clarke and Stanley Kubrick, when they created HAL, the computer from *2001: A Space Odyssey,* whose speech recognition capabilities were perfect, even extending to lip-reading? Did they think that by 2001 such machines would be commonplace or were they, as science fiction creators so often do, laying down a challenge of the future for scientists? Whatever their intention, HAL is the quintessential speech recognizer, taking verbal orders complex enough to manage a space voyage, or simple enough to play a game of chess. With HAL there is no question of continuous versus discrete speech; of large or small vocabularies; of speaker dependence or independence; of error rates; of threshold parameters; of confidence levels. There is no question of what applications for which HAL is suited. He is ready for anything. HAL is 100% across-the-board effective. HAL is fiction.

With still a couple of years to go computer scientists can already be proud of their achievements in speech recognition. For less than $100 you can buy yourself a speech recognizer that will run on your PC, recognize continuously spoken speech with vocabularies in the tens of thousands of words, improve its performance over time, learn new words, and interface with most of your software. True, it can't read lips, but research is ongoing in which lip-reading is used to assist speech recognition [1].

Hundreds of speech recognition applications are implemented and running today. Many of them are saving their owners millions of dollars. AT&T's automated call routing handles one billion calls a year and saves the company

$300,000,000 annually. For every one application area where speech recognition systems are actually implemented, there are many more where speech recognition could provide added convenience and economy, but where implementation has not yet come about. On the other hand, there are potential areas of applications—"faux apps," if you will—where managers may want to think twice before committing resources to implementation. A voice recognition system to control the brakes of a tractor-trailer, fire a gun or play the piano (now hit C#!) doesn't seem like a great idea.

6.1 Criteria for a Viable Speech Recognition Application

The general criteria for all applications of any technology revolve around convenience, safety, health, entertainment, and economy. Sometimes these forces are in opposition—certain forms of entertainment (hang gliding) compromise one's safety, certain conveniences (fast foods) compromise one's health. A positive balance must be achieved by an implemented technology if the implementation is to succeed. This is as true of electricity as of speech recognition, but these general criteria are not the purpose of this section. They need to be mentioned as general, overarching criteria, as *necessary* conditions for success. The fact that call routing is profitable for AT&T is a condition without which AT&T would not have installed the system, but by itself is not a *sufficient* condition for success. The implementation must also meet some specific criteria for speech recognition applications. Three such criteria are discussed in Sections 6.1.1, 6.1.2, and 6.1.3. They are neither independent of one another, nor exhaustive, but they indicate that more is needed to make an application viable than the fact that the technology is available and works reasonably well. To be well-received in the market place, a need must be satisfied that exceeds the possible unnaturalness and inconvenience of using an unfamiliar and imperfect technology.

6.1.1 Hands Busy, Eyes Busy

In all the activities of living we operate in three modes: we are alone, we interact with people, and/or we interact with machines.[1] We interact with people, for the most part, through speech and hearing; we interact with machines, for the most part, through hands and eyes. To improve our interaction with people we bring in underutilized channels of communication, most often the hands and eyes. When I interact with my students in class I am often talking

---

1. Animal lovers will, I hope, forgive this mild oversimplification.

and writing or gesturing, and I am hearing their responses while scanning the audience with my eyes to read their gestures and body language. To bring other channels of communication to our machine interactions, we must be able to talk to the machines (speech recognition), and listen to them, too (speech synthesis, discussed in Chapter 7).

When the hands and eyes are busy and there is still a need for further communication with the machine, then speech recognition is viable. "Hands busy, eyes busy" is intended in the very broad sense of being unable to use hands or eyes either because they are literally busy, or because other conditions such as quadriplegia, darkness, environmental conditions, etc., prevent their normal usage. Thus early applications of the technology were found in aids to people with disabilities, use in the cockpit, telecommunications, factory floors, darkrooms[2]—all scenarios in which the hands and/or eyes are unable to perform necessary functions for various reasons. On the other hand, there has been much resistance to the idea of a typewriter that takes general dictation because for the most part the hands and eyes are sufficient for typing, though it has been shown that one can dictate faster than one can type. Many applications that appear "sexy" for speech recognition—voice control of the television set—are marginal because hands and eyes are capable, even if hands and eyes are inferior in other ways such as speed or accuracy.

6.1.2 Remoteness

It is doubtful that the telephone would have been invented if people communicated by use of hands and eyes, that is, via sign language. There is nothing wrong with sign language, per se. Sign languages are fully developed languages capable of the same nuance of expression as spoken languages. They happen not to be nature's choice unless circumstances like deafness force the choice to go that way. (Perhaps if sign language were nature's choice, television would have been invented before the telephone.) With machines, in some sense, we are using sign language, at least insofar as we generally use hands and eyes. However, if the machines are remote, if they cannot be reached, or cannot be seen, then what? The easy answer is that we talk to them, even if it means calling them on the telephone.

---

2. Surely one reason human language is naturally vocal is the need to communicate in the dark. (The primary reason is undoubtedly to leave the hands free to work.) Darkness renders the eyes useless, and it is no accident that Kodak is one of the companies that pioneered efforts in speech recognition for use of speech in the darkroom and other environments—perhaps a battlefield—where no light is available.

Today most telephone communication with machines is done via the touch-tone keypad. That is most likely the way you query your telephone answering machine remotely, if you do it at all. The touch-tone signals are an extension of your fingers, and they are one way of breaching the remoteness. Eventually, however, a richer form of person-machine communication will be needed than that provided by a few beeps. Moreover, it's difficult to remember all the codes: a code to back up a message, one to erase the last message played, one to play new messages, one to play old messages, etc. Words would be easier.

During off business hours, I can suspend delivery of my local newspaper via touch tone communication with the circulation department's computer, but I must press 24 keys and listen to their verification. The process takes four minutes. When a person in the circulation department takes my order, it takes about one minute. I estimate with speech recognition, since everything has to be repeated back for verification, it would take two minutes, still a considerable saving of time. Moreover, the touch-tone system excludes customers without touch-tone service (about 35% of telephone customers in North Carolina do not have touch-tone service), so they are forced to call during business hours.

The example in the previous paragraph is a reflection of global happenings. While in the United States at large, 25% of the phones lack touch-tone service, in Europe the figure is much higher, around 75%. To the extent that it is desirable to communicate with computers at remote locations, it is desirable to have a speech recognition interface.

Another form of remoteness is created by hostile environments. When space walkers need to communicate directly with computers—and they will one day—it will have to be by voice. Keyboards, mice, track balls, touch pads, and joy sticks are difficult to manipulate in a space suit.[3]

Remoteness is a relative term. To a quadriplegic, a light switch is remote, and a voice-controlled computer to control that switch is a boon. In the dark a light switch is remote, and voice control is viable in that situation, too. People who *need* the extra help will tolerate the errors, the need for repetition, and the extra cost. However, most people, most of the time, given current technology limitations, will find voice-controlled light switches overkill and overpriced, and such an application would not find a large market of buyers.

---

3. The fact that much of what an astronaut does during a space walk involves computers, or at least computer chips, does not obviate the need for more direct communication with computers that record data, control airlocks, and so on.

6.1.3 Miniaturization

The age of the palm-top computer is upon us. Current models are capable of telecommunications, keeping your calendar of appointments, and beating you at solitaire when you don't have any appointments. Communication with these computer dwarves is through poking a tiny keyboard with a stylus, or using some kind of pointing device including touching the screen with the stylus. Sophisticated (a word of Ancient Greek origins meaning "expensive") models permit written input. Those of us who passed Ancient Greek by writing verb conjugations on our palms are well aware of how limiting that alternative is.

The palm-top will come of age when you can talk to it: "Reserve a time slot from two o'clock to three-thirty for a meeting with mister f-o-s-t-e-r-l-i-n-g on Thursday, June 25" is likely to be superior to the stylus or pointing device method of achieving the same end. On the other hand, a voice-controlled appointment calendar makes little sense on a lap-top computer where a keyboard and pointing device are already available.

Miniaturization combined with speech recognition will result in a new wave of applications such as voice-controlled hearing aids. (The ear is a perfect place to received voice commands—there are even microphones that are placed in the ear to pick up the voice.)

6.2 2001 Won't Be 2001

The nihilistic title of this section is just another way of saying what I said earlier: "HAL is fiction." No speech *recognition* system deployed anywhere today approaches even distantly the kind of speech *understanding* that humans perform on a daily basis, or that any of the fictional computers—HAL, Star Trek Computer, etc.—are depicted as performing, namely comprehension of context free, continuously spoken, idiomatic speech in noisy environments.

There are two reasons I am beating this dead horse. (I am revisiting much of what was said in Chapter 3.) First, many readers of this book may one day want to purchase a speech recognition system. It may be for personal use, use in a small business, or use in a corporate environment. It is of crucial importance to be cognizant of both the potentialities and limitations of the technology. This is especially true today, where seemingly miraculous technologies are implemented on a yearly basis. In light of the stunning successes in electronics, telecommunications, medicine, space and sea exploration, and entertainment, it's not at all implausible that a computer should equal, nay surpass, the human being in human language understanding. However, it just ain't so, and buyers need, "Just the facts, ma'am!"

The second reason is that many readers may already be in the speech processing business, in a role of manager, sales personnel, technical writer, or marketer. You, too, should be circumspect about claims for the technology. When users' levels of expectation rise to an unrealistic height, and they are subsequently disappointed, they tend to rebel by damning all the related technology. This is seen clearly in the area of automatic machine translation, where, due to excessive optimism, expectations were raised to stratospheric levels. The low level of actual results in the 1960s led the government to issue a policy of not funding such research, and progress in the area suffered a ten year setback. Nothing as extreme will happen in speech recognition, which is far more mature today than machine translation was in the 1960s, but nonetheless it is unhealthy for users to have unrealistic expectations as it may have negative consequences in the market place.

Thus, I am urging you to use common sense (and what you've learned from this book) when you read vendors' literature that claim that "now you can dictate anywhere and convert your voice tapes to text later, automatically, using your computer and continuous speech recognition software." This misleading advertising may be contrasted with another company's approach, which states "We understand the advantages as well as the limitations of automatic speech recognition and we are happy to share this knowledge with our customers."

Finally, I hope that you will avoid landing on the planet of the clueless, wherein resides an editor of a well-known speech technology publication, who wrote as an introduction to a piece about how Stephen Hawking uses speech synthesis to communicate: "Speech Recognition Enables Dr. Hawking to Explore the Universe." There is no mention at all of speech recognition in the text. Dr. Hawking is speechless, and so was I.

6.3 The Role of Human Factors in Speech Recognition Applications

Contrary to popular opinion in at least some circles, applications of speech technology don't generally fail because a speech recognizer's accuracy is 93% instead of 97% . . . They fail because human factors concerns are not addressed.

Daryle Gardner-Bonneau. preface to the *International Journal of Speech Technology*, Vol 2, No. 1, May, 1997

The biggest factor in any speech recognition application is the human factor. Speech recognition is but one of thousands (millions?) of kinds of systems in which the human being plays the primary role. Human factors comprise a field of endeavor that cuts across the fields of psychology, design, industrial engineering, and computer science, at the least. It is the study of the characteristics of human beings that are applicable to the design of systems and devices of all kinds. It strives to attain the knowledge necessary to achieve compatibility in the design of interactive systems of people, machines, and environments to ensure their effectiveness, safety, and ease of performance. An understanding of human factors is necessary for those who conceive, design, develop, manufacture, test, manage, and participate in systems.[4]

Speech recognition is a fundamentally human activity in the deepest sense: Without humans there would be no speech; without speech there would be no humans. Making computers recognize speech is a major step toward the goal of imparting human-like intelligence to a machine. To get machines to perform this very human activity, we must make our systems sensitive to human characteristics. We ourselves require the same sensitivities, but they are available to us through introspection. The machine has no such advantage. Any insight it has into how humans will interact with it must be endowed by its human designers.

Let's look at a simple application to see the difference between a human factor and a machine factor. Suppose we want a speech recognizer to distinguish letters of the American alphabet so that any word could be recognized through spelling. An important machine factor is the similarity in pronunciation of the letters *b, c, d, e, g, p, t, v,* and *z,* which are easily confused and lead to errors. A solution to this problem is to use a word-based alphabet such as pilots use. Instead of *b,* one says *bravo;* instead of *p,* one says *papa,* and so on. *Bravo* and *papa* are less likely to be confused than *b* and *p.* However, there are human factors. Humans know their alphabet, but most humans would have to learn the pilot's alphabet. Moreover, it takes a human longer to say *bravo* than *b.* Thus the application could not be implemented without training the user, and it would perform more slowly than the letter-based system.

A more subtle example of human factors at work is found in template-based discrete-word recognition. It is one that has been "discovered" numerous times: The likelihood of a recognition error is higher if the previous utterance was in error. At first blush, one might think the opposite. After being in error, the user would tend to be more careful. In fact, that is the problem. When human speech is misunderstood, the human repeats it more emphatically, and thus the word's pronunciation deviates even more from the pre-stored template.

---

4. These definitions are extracted from the homepage of the Human Factors and Ergonomics Society. The URL is http://hfes.org/~hfes/About/Menu.html.

As human frustration with errors grows, pronunciation deviates more, errors abound, frustration increases . . .

Once this aspect of human behavior is understood, the user may be instructed to react calmly to errors and focus on replicating the utterances of the training session. If the input is continuous speech, where the same phenomenon is also observed, the user may be instructed to continue speaking naturally but to avoid "running words together," that is, to reduce the degree of interword coarticulation—a chief source of error in continuous speech recognition.

It is impossible in a book of this size to accomplish more than a superficial scratch of the surface of the role of human factors in speech recognition. A small claw mark may be found in the section entitled *Human Factors* in Chapter 3. There I observed what I consider to be the fundamental axiom of the human factors of speech recognition, which I repeat here in somewhat different words: User cooperation is necessary for a successful speech recognition application. Two excellent reports on human factors in speech recognition may be found in [2, 3].

6.4 Application Areas

Applications in speech recognition may be viewed on a sliding scale. At one end of the scale are applications that are *real, implemented,* and *commercially successful;* at the other end are applications that are *hypothetical*—somewhat analogous to Einstein's "thought experiments." In the middle of this sliding scale are applications that are real and implemented, but not commercial, such as one might build for personal use. The scale is a continuum because the notion of "commercial" is imprecise. Even the notion of "hypothetical" comes in degrees. A speech recognition interface to a microwave oven is less hypothetical than a speech recognition interface to a hearing aid because the former is so similar to many implemented interfaces, but the latter is definitely futuristic.

Applications may also be broadly classified according to principle realm of impact. This taxonomy also has its faults—the areas both overlap and "underlap" (i.e., some applications may defy the classification)—but any degree of organization is better than none. In this vein I would like to propose five general areas of speech recognition applications: *assistive technology, telecommunications, command and control, data entry and retrieval,* and *education.* They overlap in the sense that a speech recognition interface to a wheel chair is both assistive technology and command and control—we have to focus on the primary purpose of the interface. They underlap in that there are some unique applications that do not seem to fit any category such as a speech recognition front-end to a palm top computer that translates English into a foreign language. Nonetheless we press boldly on. Sections 6.4.1–6.4.5 discuss various applications within their category.

6.4.1 Assistive Technology

Assistive technology is the use of technology for the primary purpose of aiding persons with disabilities. This chapter focuses on aids through speech recognition.

Disabilities, like so many concepts in this book, defy any kind of discrete classification. On the one extreme are people like Stephen Hawking, deprived by disease of all movement and speech; and there are the milder but significant (especially if they're yours) disabilities that accompany normal aging. Clearly, there is much ground in between, and it is doubtful that we could classify disabilities on any kind of a linear scale, nor should we try.

The Boeing Company has a history of interest in speech recognition for use in aircraft. However, it was also, perhaps unwittingly, instrumental in the use of speech recognition for the seriously disabled. Some 20 years ago one of its programmers was rendered quadriplegic from a boating accident. The group working on speech recognition had several discrete-speech, speaker-dependent recognizers that were unused at the time, so they developed a voice-controlled interface that allowed the disabled programmer to operate his computer, control a robotic arm for fetching manuals and turning their pages, and interface to a FORTRAN development system so he could continue to work. The system eventually evolved into a robotic vocational workstation for the physically disabled professional. (See [4] for more details.)

Since then numerous such applications have been developed for providing computer-controlled assistance with speech interfaces to persons deprived of their full faculties because of an accident or disease. In a properly designed home, voice can be used to control light switches, appliances, doors, windows, plumbing fixtures, etc. Speech recognition can also enable handicapped individuals in the workplace by allowing a voice interface to a computer, and controlling telephones, fax machines, lights, etc.

6.4.1.1 Universal Design and "Curb Cuts"

Some of the best ideas arise accidentally. The most notable instance is depicted by Isaac Newton getting bonked on the head by a falling apple and coming up with gravity (double meaning intended). Some years ago the designers of our streets and sidewalks implemented "curb cuts," portions of the curb that join the street smoothly, for the purpose of accommodating wheelchairs. However, wheelchair users were not the only beneficiaries. In fact, by pure count, they were in a distinct minority of individuals who enjoyed the benefits of curb cuts, which included people with canes, people pushing baby carriages, skaters, cyclists, kids pulling wagons, boy scouts piloting little old ladies, and on and on. Out of this piece of serendipity arose the concept of **universal design**—designing technology both as assistive for the impaired and as useful for the unimpaired. Curb cuts became symbolic of the universal design concept.

The concept has made its way into speech recognition. Let's focus on the home thermostat. At first blush a voice-controlled thermostat sounds a bit extravagant, something for that private compound on Malibu beach. Still, try to both read and adjust your thermostat while sitting in a chair. In many homes, for many people, this isn't possible. So a speech recognition interface would be assistive for many impaired individuals, including the bedridden and the visually impaired in addition to wheelchair confinees. For the unimpaired imagine a telephone speech recognition interface to that same thermostat. Before leaving home for work you can set the thermostat to a thrifty but uncomfortable level. Rather than face coming home to a hot or cold house at the end of the day, you call your thermostat and set it to a comfortable level for your arrival home. It makes sense, then, for thermostats to be designed universally with voice control—everyone benefits. This scenario can be extended to many household functions, from setting burglar alarms to adjusting the oven. These are "curb cuts."

Voice activated telephone dialing was at one time an assistive technology. Then came car phones followed hard upon by car pileups as drivers dialed and drove their cars into one another. Then came the voice-activated car phone. Several states are considering making voice activation mandatory in car phones. Extremists want to remove the keypad entirely. Yesterday I saw a telephone with speech recognition dialing for sale in a pharmacy, as common as aspirin. This is an example of a curb cut that is also a real, implemented, commercially successful speech recognition application.

Another example of a curb cut finds its roots in the Boeing programmer who used speech recognition to write FORTRAN programs. In today's terms we would think of it as an interface to a workstation. People with all manner of disabilities—from victims of carpal tunnel syndrome to the catastrophically paralyzed—need speech recognition interfaces to computers, and there are several such products available to them.

Unimpaired persons, too, can benefit from a voice interface to a workstation. Xspeak was an MIT Media Lab project that used voice input to replace the mouse (or other pointing devices) for switching among applications under a Windows-style operating system [5]. It required only a discrete-word, speaker-dependent small-vocabulary recognition system. By speaking the name of a window, that window could be brought into focus, even if it was invisible on the desktop. Other commands allowed the user to raise or lower any window containing the cursor.

For any user, impaired or unimpaired, Xspeak was useful because it saved the trouble of locating and manipulating the pointing device, an inconvenience when the user's hands and eyes are busy on the keyboard and screen. It was especially useful for resurrecting buried windows. (These were in the days preceding modern windowing systems that avoid this problem, but the illustration is nonetheless valid.)

To measure the utility of this particular application, a small number of student programmers used Xspeak over a course of several months. In [6] it is reported that ". . . a majority of the users were able to use Xspeak effectively and incorporated it into their daily work routines. Some were extremely enthusiastic and competed for access to Xspeak."

Many hypothetical universal design applications of speech recognition are easily thought up, and many of them are discussed in the literature. For example, certain impairments prevent a person from voting in the ordinary way. They are allowed to take someone with them into the voting booth, but this compromises their privacy. A speaker recognition interface to the voting machine would help these persons. Ah, but now that we have it, let's also use it for dyslexic, blind, or illiterate voters, and once such a system is in place, voting by telephone becomes an option that may improve the dismal voter turnouts that have plagued the U.S. electoral system for years.

Here is another hypothetical but realizable scenario. A marvelous aid to the hearing-impaired would be a speech-to-text machine, essentially speech recognition carried to its final destiny. A deaf person could use this for telephone communication, and if the device were portable, for all communication. With such technology in place it now becomes a simple matter to close caption television programs automatically. (It's done stenographically nowadays, in case you're wondering.) Now the benefit extends beyond the hearing-impaired to locations where a television picture is available but there is too much ambient noise to hear the program, such as a sports bar or gym. Similarly, why not use the device to enter speeches into the Congressional Record? That way no one, not even a poor stenographer, need suffer through windy rhetoric "for the record." While we're at it, let's use the device to produce transcripts of court proceedings, of business meetings, of lectures, and so on.

A speech recognition interface to a vending machine seems as though it might be specific to assistive technology (it would include some billing method). However, the advantage of having less moving parts to fail such as stuck buttons and trapped coins translates to a more reliable, hence more profitable operation, which is good for business. It also results in a coterie of dedicated customers, a more compact machine, a less breakable machine, and less broken hearts when that chocolate treat you looked forward to all afternoon is not imprisoned in a lifeless showcase, so close yet so far away.

6.4.2 Telecommunications

We need to distinguish between speech technology *for* telecommunications, and speech recognition *by* telecommunications. For example, a voice-controlled dialer is used for telecommunications because its primary purpose is to facilitate the use of a telephone or fax machine. Contrariwise, the primary purpose of voice access to a bank account by telephone is fiduciary; the use of the telephone is incidental. This section is about speech recognition applications used for telecommunications.

The use of speech recognition in telephone call processing is beginning to leave its infancy. For years it was kept in its crib by the challenge of high accuracy, speaker-independent recognition over phone lines in possibly noisy environments. These obstacles have been partially overcome by broadly trained HMMs for word-spotting recognizers with very small vocabularies. (For example, if a caller says, "I want to place a collect call," the system "spots" the word *collect,* and routes the call appropriately.")

Similarly—and becoming commonplace—when a collect call is made to a location, the person answering the telephone is asked, using a combination of synthetic voice and a voice recording of the name of the person placing the call, if he or she wishes to accept the call. The response is scanned for "yes" or "no," so if the person called says "yes, I do" it will be detected as an affirmation. The system is also capable of interpreting *yeah, yuh, yup, naw, nah, nope,* and additionally, in Canada, the French *oui, ouais, ouen, non, nen, nan.*

The next step in the maturation of this technology is the robust recognition of digits. This will permit the voice entry of telephone and credit card numbers, still done by keypad in most environments. The various phone companies all have speaker-independent recognition systems that do very well on digits, but even an Ivory Soap accuracy of 99.44% per digit, well beyond what is currently achievable under actual circumstances, will lead to an error one time in seven, too high a rate for heavy commercial use.[5]

Speech recognition interfaces to directory assistance services are becoming available. (They are out now in a few special markets for testing and evaluation purposes.) No speech recognition system yet is able to handle directory assistance in general simply because there are too many proper names to train for a word-level recognizer and too many different spellings for a given phonetic pronunciation even assuming (falsely) that a phonetic based recognizer would be sufficiently accurate.

---

5. Assuming a 16-digit credit card number and an 11-digit phone call, to complete an error-free call, one must speak 27 digits without error. The overall accuracy is (.9944) raised to the 27th power, or about 6/7.

The applications available today are of two types. One recognizes the names of cities for the purpose of routing the call to the appropriate operator. The second type is oriented to commercial directory assistance. It is a word spotting system that can handle a call such as "Is there an Indian restaurant downtown?" It spots words such as *Indian, restaurant,* and *downtown,* applies some artificial intelligence, and determines that the caller would like the phone number(s) of Indian restaurants restricted to a certain location.

Global telecommunication companies are working very hard on automated spoken language translation systems. None exist as of this writing, but the need is present as illustrated by AT&T's "Language Line." Recognizing the value of interpretation as a new way to meet customer needs and increase the demand for domestic and international minutes on its Worldwide Intelligent Network, AT&T provides on-demand interpretation for up to 140 languages, 24 hours a day, seven days a week, to clients such as businesses, emergency services, government agencies and individuals.[6] The interpretation service uses English as the base language, with an interpreter converting a series of messages from English into another language and then from that language back into English. The service enables people speaking different languages to communicate immediately and effectively across language and cultural barriers. Some of the industries that take advantage of the interpretation service are law enforcement and emergency services, health care, insurance, finance, utilities, telecommunications, government agencies, and transportation.

Language Line is a human-driven service, but we can envision the day when computers take over the chore. As with all complex technologies, achievements are likely to come in increments. High-quality translation between written forms of languages is still an unachieved goal. The interested reader is urged to review the excellent summation of the state of the art of machine translation of written texts at URL: http://www.he.net/~hedden/intro_mt.html. Interpretation of spoken language can be viewed as sandwiching a machine translation system between a speech-to-text system on the one end, and a text-to-speech system on the other end, as suggested in Figure 6.1.

The final application area of speech recognition for telecommunications is automated service requests. Telephone companies take millions of service calls a year within the very limited context of service start-ups, service discontinuations, and repair service. The front-end of the application would ask callers to state which service was desired (and, as always, give them a chance to connect to a live operator, wherein they would be made to wait on the line for

---

6. More details such as the actual languages, are available at the URL http://www.att.com/languageline/.

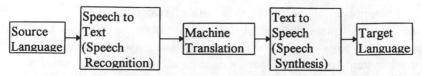

Figure 6.1 An automatic language interpretation system.

half-an-hour to give them a higher appreciation of the automated service). For a discontinuation of service, the caller would state the telephone number and the desired date of cessation. To keep the process entirely automated, users will have to give a personal identification number, issued in advance, to prevent malicious tampering with their phone service.

Service calls would be handled similarly. The user would state the phone number of the troubled phone line. Since most service calls are for "dead" lines (i.e., no dial tone), only the phone number is needed. Other service calls, say for "cross talk," would be handled by offering the user a menu of choices, and having them speak the relevant choice.

Service start-ups would be the most challenging as the caller would have to transmit names and addresses to the speech recognizer, in many cases resorting to spelling. When (not *if*) this service is made available, it is likely to begin with cessation-of-service requests and repair service requests. When systems become more sophisticated, service start-ups will become viable.

6.4.3   Command and Control

The expression "command and control" is, as you might guess, borrowed from military parlance. It refers to central locations where officers in charge of an operation can issue commands to control the movement and deployment of men and machines. In speech technology the expression refers to the control of machines via computers which are themselves instructed through speech.

Many of the applications we have already looked at are command and control type applications, but their primary purpose is assistive and they were discussed previously. For example, the control of the household or office environment through verbal instructions is primarily assistive, but is also command and control.

I'd like to look at command and control applications in four environments: industrial, military, healthcare, and hostile.

6.4.3.1   Industrial Environments

Factory floors teem with hands busy, eyes busy activity. Many early applications of speech recognition are found in such environments. One such application involves the inspection of micro-electronic parts under a microscope, as they are advanced by a belt. The eyes are busy focusing through the microscope, the hands are busy manipulating the part. Speech recognition is used to control the movement of the part between the main conveyor belt and the inspection station. It may also be used to adjust the focus and lighting on the microscope. An added benefit of speech recognition is that any flaws that are detected may be entered directly into a computer by voice, obviating the need to remove both hands and eyes from the task to record written data.

Workers in factories often have their hands full and need to use their eyes to avoid accidents. Often they have a need to open and close doors, adjust lighting, adjust ventilation, set timers, etc. Traditionally they set their burden down to accomplish the manifold environment adjustments necessary in the course of a day's work. Voice control provides the means to work more efficiently. There is also a safety benefit from controlling heavy machinery by voice: You needn't get as close to it as you do in a hands-on operation, reducing the risk of an unpleasant interaction between the moving parts of the machine and a flap of clothing.

I intend the term "factory floor" to be used in the broad sense of a location where work involving machines and/or manual labor is being done. A post office—the back part, not the sanitized lobby that most of us know—is very much a factory floor in that sense. In 1980 one of the first commercial speech recognition systems ever to be deployed was set up in a central post office in New Jersey. Packages were routed by speaking their zip codes into a microphone thus controlling the conveyor belts carrying the packages. The speech recognizer was a discrete-word recognizer trained to the digits.

Another "factory floor" is the tarmac where airplanes park to load and unload passengers and their luggage. Conveyor belts controlled via speech recognition were set up by United Airlines at Chicago's O'Hare field to assist workers, who could route luggage by calling out their airport codes—DCA, LAX, DFW. Noise management was a major challenge to the success of this operation and was achieved by the use of close-talking microphones, which pick up sounds only within a small radius around the microphone. Both accuracy of luggage routing and number of pieces handled per minute increased through the use of the voice system.

6.4.3.2 Military Environments

We already examined in considerable detail a hypothetical command and control application of speech recognition to manage tank warfare (see Table 3.4). A similar system for the input of voice commands to an *Airborne Warning and Control System* (AWACS) aircraft has also been developed [7]. The military has been the most active supporter of speech recognition research in this country and quite possibly the most prolific user. Speech recognition systems have been employed by the military in applications ranging from assisting in the repair of tank engines, to accomplishing minor tasks in the cockpit such as adjusting radio frequencies.

The cockpit of a modern military aircraft, both fixed and rotary wing (helicopter), is a busy place for the hands and the eyes. Moreover, many of the aircraft systems are too complex to be operated by humans alone, and require the use of computers. The computers, however, are subject to human control and may be instructed through use of a keyboard or touch screen. Manual input strains even further the task load on the hands and eyes. It is a perfect scenario for speech recognition, and indeed, researchers at Wright-Patterson Air Force Base, Fort Ord, Ames Research Center at Moffett Field, and the Aberdeen Proving Ground have been studying how to integrate voice into the command and control needs of the cockpit.

Experimental systems have been built and tested for voice-controlled radio tuners, navigation aids, target acquisition systems, and threat-avoidance systems. Under ideal circumstances, the voice systems integrate well with other cockpit activity, but conditions in a war plane are never ideal. Pilots may be required to operate their aircraft at high speeds close to the ground, where a wrong decision may lead to catastrophe. They often fly at night and under adverse weather circumstances. The cockpit environment is harsh from the point of view of speech recognition. It is noisy, hot, and full of vibrations. The users are under the psychological and physiological factors of stress, fear, and fatigue. Moreover, they may or may not be wearing masks, which affect their voice quality. All of this conspires to lower speech recognition performance. An avoidance system with voice input that works well in a simulated attack may fail in an actual attack, where the pilot is truly afraid, and the fear causes voice alterations.

Advances in microphone technology and increases in the robustness of speech recognition systems have made the use of voice control in the cockpit viable. Nonetheless, one finds such remarks in the literature as ". . . merely adding voice technology to existing displays, or trying to replace visual/motor displays with voice technology on a one-to-one basis can create problems for the pilots." One is reminded of the final scenes of the motion picture *Star Wars,* where pilot Luke Skywalker eschews his computer-controlled weapon system and stakes the fate of the Galaxy on himself and "The Force." A good reference for the issues of voice in the cockpit is [8].

6.4.3.3　Medical Environments

The most common deployed application of speech recognition in the healthcare industry is in data entry and report generation, to be discussed in Section 6.4.4. I put this section here because it is my belief that speech recognition should be, and will be, used for command and control in the hospital room, the operating room, and the medical laboratory.

Most people who reside in a hospital room are likely to be physically and/or psychologically impaired. Many of the applications of speech recognition in the assistive realm could be used with great benefit in the hospital room. These include television control, bed adjustment, water and ice dispensing, light and door control, voiced-activated call button, and so on. Speaker-independent recognition would be required, but the vocabulary could be small and discrete utterances would suffice. The patient could be taught the commands needed during pre-operation prepping, where patients generally have excessive time on their hands anyway.

A surgeon in action is a stereotypic instance of hands busy, eyes busy. In the operating room she is surrounded by support staff who control lighting, record information, adjust the operating table, etc. Since much surgery today is conducted under a microscope, that instrument must somehow be adjusted when necessary, most likely by the surgeon since it's her eyes that must be accommodated. Some of these functions could be controlled by voice, putting less personnel in the operating room, with a concomitant cost saving.

The Zeiss company has experimented with a voice-controlled microscope to be used in ophthalmic surgery, but as of this writing such instruments are rare in practice. "Puff" control microscopes, on the other hand, are commonly found in the operating room. The surgeon blows puffs of air to control the microscope parameters. (Such devices are also in use for severely disabled persons.) They could be considered precursors to the voice-operated microscopes that will undoubtedly become common in the next few years, thanks to the recent advances in speech recognition technology.

In a hospital laboratory, or any laboratory where chemicals are handled and sterile conditions are required, technicians find themselves needing "a third hand" to start an exhaust fan, turn on a light, open a door, start or stop a machine, set a timer, and so forth. That third hand could be the vocal cords.

6.4.3.4 Hostile Environments

Like all frontiers, space, the final frontier, is hostile. The need for computer-controlled machinery is well illustrated by astronaut Jim Bowman's commanding HAL in *2001, A Space Odyssey* to "open the pod bay doors." The utility of being able to control computers by voice is nowhere more apparent than when you're in outer space ensconced in a space suit.

The inside of a space craft or space station—most of us are familiar with that scene nowadays—is an intensely busy, cluttered, awkward place to work, where the hands, in addition to traditional activities, are often used just to hold one steady. Although to my knowledge voice-controlled computers are not presently used in space, I predict that as the technology becomes more commonplace on Earth, it, too, will migrate toward the stars.

Other hostile environments include locations near dangerous chemicals or nuclear material, dirty or dusty areas like the paint shop of a car manufacturer, dark or inaccessible areas such as furnaces, confined spaces such as airplane or tank engines, the vicinity of high tension wires, and heights such as spires. Kelway [9] has suggested a number of potential command and control applications of speech recognition in the inspection of power cables in and around electricity generating stations such as the vocal adjustment of machinery. Speech recognition could offer direct contact to computers at the same time as allowing operators to climb pylons or work in holes.

6.4.4 Data Entry and Retrieval

Data entry is altering the contents of a computer's data file; data retrieval is discovering something about the contents of a computer's data file. These operations can be accomplished by voice when the "hands busy, eyes busy" or "remoteness" conditions prevail. I am using *data entry* in the broad sense that includes completing application forms, filling a table with numerical data, and using a word processor to create a document.

6.4.4.1 Dictation Systems

Fifteen years ago the "listening typewriter" was a hypothetical application [10]. Today dictation systems have vocabularies as large 64,000 words[7] and are capable of transliterating speech at well over 100 words per minute. The degree of accuracy depends on many factors, including the user's skill with the system, how much the system is used (it learns as it goes), and the vocabulary to be spoken. Microphone placement and degree of ambient noise also play a role in accuracy. The systems have moved from hypothetical to real, implemented, and commercially successful.

Early commercial dictation systems worked best with discretely spoken speech. The attempt to use connected speech often led to unacceptably high error rates. Because it is tedious to speak one word at a time, these systems didn't find many areas of application. Most people would prefer to type, if they could, than dictate in this manner. In one area, however, the need was so great that the difficulties of speaking a word or phrase at a time did not dis-

---

7. *PCWeek Online* of June 18, 1998, (URL: http://www.zdnet.com/pcweek/news/0615/18espeak.html) announced that IBM's *Via Voice*® dictation system had been expanded to a vocabulary of 64,000 words.

courage application. That area was medical reporting, and in particular, radiologist reports.

Radiologists are physicians responsible for interpreting X-rays, MRIs, and CAT scans and reporting their findings. The usual method is for them to speak into a tape recorder. Later, a transcriptionist types up the report, which is returned to the radiologist for final approval, and then entered into the patient's medical record. The process is expensive, costing hundreds of millions of dollars per year, and inconvenient for everyone: The radiologist must dictate, proofread, and sign off; the attending physician—the one who orders the X-ray—must wait on the radiologist's report; and the patient, too, must wait for the completion of the process, possibly on tenterhooks when a brain tumor is the object of interest.

The ability to dictate directly into the computer results in large cost savings. Moreover, the radiologist both dictates and reads the report at the same time, so when the report is completed, it can be signed off right then and there. The report can be entered into the patient's medical record electronically and be immediately accessible to all interested parties. Indeed, most reports need never appear on paper, nor should they have to.

Only the inconvenience and inaccuracy of early speech recognition systems held back its wide usage. Radiologists refused to invest in systems that required them to speak unnaturally and cope with errors. Nowadays accurate continuous recognition is making reporting more palatable, and the cost savings alone is a strong motivator. Adaptable systems adjust the speech models to accommodate each individual speaker, and the use of artificial intelligence permits systems to distinguish among words and symbols that sound alike, such as *to, too,* and *two;* or the "colon" as a body part and the "colon" (:) as a punctuation mark.

Performance is also enhanced by the use of "macros"—computer science jargon for "short cus." Most medical reporting has certain standard sentence forms that occur frequently, such as "the patient was seen by me on &date," where the "&date" means "date to be filled in." The system can be set up so that when the physician says "Seen December 20, 1998," the system automatically writes "the patient was seen by me on December 20, 1998."

The talker must still behave like a "sheep." No system will deal well with mumbling, slurring, multiple hesitations, stuttering, talking while eating, and so on. The human factor is *always* there. With today's systems the demand on the user is minimal, and it's common for the positive factors of dictating to outweigh the inconvenience of learning new reporting habits.

By the end of 1998 IBM's MedSpeak/Radiology continuous speech recognition product had been deployed with tremendous success in numerous medical centers. Some hospitals are now using MedSpeak/Radiology exclusively for reports, and the other nonradiologist physicians, in hospitals where the product is available but not mandated for use, actually request that radiologists prepare their reports using the system.

Dictation systems of various kinds are so prevalent today that books are written about them. A good introduction to the subject is found in [11].

6.4.4.2 The Dental Chair

Have you ever had a periodontal examination? The periodontist sticks his head and both hands in your mouth and probes the bone depth around each tooth surface, calling out tooth numbers, names of surfaces and bone depth in millimeters to an assistant. Lacking an assistant, he frequently retracts his body parts to record data, then reenters. Here is a quintessential hands busy, eyes busy, scenario that cries out for speech recognition. The technology is in place, but to my knowledge this application remains hypothetical.

6.4.4.3 Travel Information

In both the United States and the United Kingdom, systems have been implemented that allow travelers to call up a computer, receive travel information, make reservations, and purchase tickets. In the United States, the emphasis is on air travel; in the United Kingdom it is on rail service. The systems are not yet widely used commercially. They are midway between hypothetical and commercially successful.

The travel information systems involve not only speech recognition on the front end, but a form of artificial intelligence called *planning*. In essence, the system must have some degree of understanding when you request "information about morning flights from Atlanta to Dallas." Based on that understanding it plans out the most useful answer it can compute. The system has some leeway as to how to respond. It may decide to give only nonstop flights, or nonstop and direct flights; or it may give all combinations involving only a single stop. Alternatively, it could respond with a question: "First class or coach?" Furthermore, it could ask the traveler for a price limit on the ticket before presenting a choice of flights or ask the travel whether commuter flights should be included, and so on.

The system must also be smart enough to remember the context of the transaction, so after answering questions about the flight from Atlanta to Dallas, if the traveler says "What about to Washington?" the system must take this as an inquiry about morning flights from Atlanta to Washington.

One airline is using a voice-driven system for its employees to schedule their flights. (Employees are presumably more tolerant and cooperative than the general public.) The system is intended for deployment for corporations and the general public in 1999.

In the United Kingdom, a similar spoken language system exists for rail travel, called *RailTel*. The continuous speech recognizer has a recognition vocabulary of 1,500 words, including 600 station names. The recognizer is adapted to deal with speaker-independent telephone quality speech. Prior to deployment the system was tested by having test subjects interact with the system in a realistic manner. About three-quarters of the calls were successfully completed. As of this writing the system is still considered experimental.

6.4.4.4 Banking Systems

Many banks now permit you to access account information by telephone, entering data via touch tones and receiving information via synthetic voice. Speech recognition is desirable where touch tones are not available, which is 25% of U.S. households and much larger percentages in Europe and Japan.

For a number of years the Japanese banking system has used a system called ANSER, which permits customers to phone up and receive accounts information. A typical interaction goes like this:

Bank: Hello, this is your bank telephone service center. What is your account number?

Customer: five, five, . . .

Bank: What is your secret number?

Customer: one, two, . . .

Bank: What service do you wish?

Customer: I would like an account balance.

Bank: Your current balance is 201,234 yen. If you would like to have your balance repeated, please say "once more." If not, say "okay."

Customer: Okay

Bank: Thank you very much.

The system, which is installed throughout the country, is said to handle several hundred thousand calls daily [12]. Similar systems are being tested in the United States, and several are already in commercial use, generally operating side-by-side with a touch-tone system, so the user is given a choice of using the touch-tone key pad or speaking to enter the account number and PIN number and to make particular requests.

Such a system is currently undergoing testing in at Lucent Technologies Inc. [13]. The system is intended to recognize such spoken requests as "Could you read today's balance on that credit card?" and "I want to talk with a customer service representative," in addition to digits. System testing resulted in a 10% error rate, too high for commercial application, but encouraging enough to work toward an improved implementation. Look for these voice systems to spread at the turn of the millennium.

6.4.4.5   Stock Quote Systems and More

Systems are currently deployed that allow callers to use speech recognition to get stock and option price quotes by telephone. The caller can ask for any of about 14,000 stocks or indexes by saying their name.

The world's largest package-shipping company is now using speech recognition to give customers the status of their package when they call and say the tracking number—16 letters and numbers. The system is said to have paid for itself in several months.

Customers of major catalogue companies may now call them up and use speech recognition to order items from catalogues.

One last hypothetical application is drive-through ordering of fast food—hold the pickles.

6.4.5   Education

The benefits of computers to education is unquestioned. *Computer-aided instruction* (CAI) is one of the most venerable application areas in computer science [14]. Today's powerful, inexpensive hardware and well-designed software, together with the need to provide drill and practice to reinforce learning, have permitted CAI to penetrate the classroom as never before. CAI is used with children and adults in the traditional school environment and with the work force in the corporate environment. Properly designed CAI systems provide an opportunity for anyone to acquire new skills or upgrade old ones and to do so in a cost effective, efficient manner.

Is there a role for speech recognition in CAI? Certainly "yes" for persons unable to use the keyboard or a pointing device, and again "yes" for acquiring oral skills as in learning a foreign language, overcoming a speech deficiency, or learning to read.

6.4.5.1   Language Learning

For many who wish to learn to speak a foreign language, proper pronunciation is a daunting challenge. No matter how well you master the language, if good pronunciation is lacking, that is, if you speak with a heavy "foreign" accent, people will not understand you. There are two ways to meet this challenge: one is an understanding of the science of phonetics, as outlined in Chapter 1; two is practice, practice, practice.

The computer is ideally suited to meet the second challenge. The student may be presented with a written word or phrase on the screen and asked to pronounce it. Her pronunciation is compared with a previously stored template of an ideal pronunciation. The degree of difference is fed back to the student, who may be invited to try again, or who may be asked to move on to the next exercise. Subsequent attempts at pronunciation may be prompted by having the computer speak the intended utterance for the student to imitate. There are many ways to structure such drills. One possibility is shown in Figure 6.2.

Within limitations, the computer with speech ability is the perfect language tutor. It is infinitely patient, cheerful, and positive; it can be personalized to each student's particular needs; it gives its full attention and provides feedback to every student response; it allows students to move at their own pace; and it is the incarnation of consistency. One rather severe limitation is that speech recognition technology is still not very good at deciding whether a given pronunciation is sufficiently correct. The difference between language spoken natively and language spoken with an accent is very small and quite subtle. The application as described in this section has not yet reached the real, implemented, commercially successful stage. Look for it in the next century.

6.4.5.2 Speech Therapy

Millions of Americans have impaired speech. The causes are innumerable. Chief among them are deafness, brain injury or disease, and organic disorders in the vocal tract.

Most individuals with speech disorders can learn to improve their articulation through the proper kind of practice, as determined by a speech therapist. Much of the practice, however, need not take place in the presence of the therapist. A computer with speech recognition can assist the patient with consistent repetition and feedback therapy.

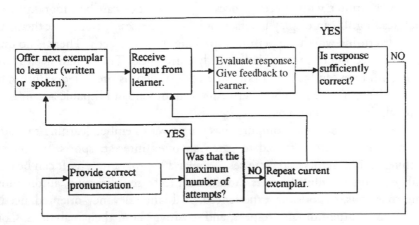

Figure 6.2 An drill system for pronunciation.

The method is similar to that described for language training. The speech therapist programs the computer to present various stimuli to the patient, eliciting either specific sounds or words that improved articulation. The patient attempts a pronunciation, which is evaluated by the computer, and returned to the patient as feedback. The application can be couched as a computer game for children, and prompts may be offered orally for illiterate individuals or to provide a target for pronunciation.

An additional advantage of computer usage is that the therapist has a complete history of the patient's progress, including all attempted utterances, the corrections, and the repetitions. Properly organized and indexed in digital form, it provides the therapist with records far more accessible than if they were in the form of tape recordings or written notes. A more detailed discussion of this kind of application may be found in [15].

6.4.5.3   Reading Instruction

Millions of Americans do not know how to read . . . at any level. Forty million are unable to read above the fifth grade level. It's shocking! So what can we do about it?

1. Make jokes about it, being sure to involve your least favorite political figure;
2. Blame something, someone, anyone: the schools, television, sex, your least favorite political figure;
3. Do something about it.

I will leave (1) and (2) to the pundits. Let us consider (3).

A computer with speech processing capabilities can be programmed to teach reading. It can display letters, words, and phrases. It can speak them, and it can determine whether a person is speaking them correctly. There is no question of perfection of pronunciation, only correctness. The system need merely distinguish between [kat] and [kit] when the student is presented with *cat*, but whether it's spoken with a Southern American or Cuban Hispanic accent is immaterial. Current technology is adequate.

A modern personal computer has the power to embed learning to read in a variety of formats designed to catch anyone's interest: sports, war games, clothing design, actors and actresses, popular TV programs, etc. It can be especially effective for adult learners. Adults tend to be self-directing, autonomous, time-conscious, responsible, experienced, goal- and relevancy-oriented, needful of individual attention and respect, and sensitive about their inabilities. Computers are perfect teachers for such students.

This technology has reached the real, implemented commercially successful stage. A variety of learning-to-read products are on the market. Check out this URL for the description of one such product: http://www.whizjr.com/products/Re100249.html

There are also systems that presently are not commercially available but are being tested in the schools. One such system was developed at Carnegie Mellon University and is described at URL: http://focus.andrew.cmu.edu/Apr97/speech.html

A second such system, sponsored by IBM, is described on the Internet as follows:

> IBM is working with one of our grantees, the School district of Philadelphia, to develop Watch Me Read, an innovative speech recognition program that helps emerging readers and children with limited English proficiency learn to read. IBM also is partnering with the San Francisco Unified School District to design an online case management system that will prevent unnecessary placements in the special education system and better address the specific academic or behavioral problems of at-risk children.

More information is available at URL: http://www.empl.ibm.com/diverse/educopps.htm

References

[1] Garcia, O., A. Goldschen, and E. Petajan, "Continuous Automatic Speech Recognition by Lipreading," in *Motion-Based Recognition,* Shah, M. and Jain, R (eds.), Norwell, MA: Kluwer Academic Publishers, Computational Imagery and Vision Series, Vol. 9, 1997, Ch. 14.

[2] Gardner-Bonneau, D. J., "Human Factors in Interactive Voice Response Applications: 'Common Sense' is an Uncommon Commodity," *Journal of the American Voice I/O Society,* Vol. 12, July, 1992, pp. 1–12.

[3] Nusbaum, H. D., and J. DeGroot, "Speech Recognition Systems," *Journal of the American Voice I/O Society,* Vol. 15, March, 1994, pp. 35–61.

[4] Fu, C., "An Independent Workstation for a Quadriplegic," In *International Exchange of Experts and Information in Rehabilitation, Monograph #37,* Foulds, R. (ed.), World Rehabilitation Fund, New York, 1986, pp. 42–44.

[5] Schmandt, C., M. S. Ackerman, and D. Hindus, "Augmenting a Window System with Speech Input," *IEEE Computer,* Vol. 23, No. 8, 1990, pp. 50–56.

[6] Schmandt, C., *Voice Communication with Computers,* New York: Van Nostrand Reinhold, 1994.

[7] Salisbury, M. W., J. H. Hendrickson, T. L. Lammers, C. Fu, and S. A. Moody, "Talk and Draw: Bundling Speech and Graphics," *IEEE Computer,* Vol. 23, No. 8, 1990, pp. 59–65.

[8] Voorhees, J. W., and N. M. Bucher, "The Integration of Voice and Visual Displays for Aviation Systems," *Journal of the American Voice Input/Output Society,* Vol. 1, No. 1, June, 1984, pp. 48–60.

[9] Kelway, P. S., "Effective Human-Machine Interfaces for Use in Industry and Commerce," *Proceedings of the European Conference on Speech Technology,* vol. 2, 1987, pp. 206–209.

[10] Gould, J. D., J. Conti, and T. Hovanyecz, "Composing Letters with a Simulated Listening Typewriter," *Communications of the ACM,* Vol. 26, No. 4, 1983, pp. 295–308.

[11] Schindler, E., *The Computer Speech Book,* Chestnut Hill, MA: Academic Press, Inc., 1996.

[12] Nakatsu, R., "ANSER: An Application of Speech Technology to the Japanese Banking Industry," *IEEE Computer,* Vol. 23, No. 8, 1990, pp. 43–48.

[13] Mikkilineni, R. P., E. L. Rissanen, and W. R. Belfield, "Natural Language Interface—An Application to Credit Card Services," *Proceedings of the 1997 American Voice Input/Output Society,* San Jose, CA: AVIOS, 1997, pp. 139–143.

[14] Kearsley, G., B. Hunter, and R. J. Seidel, "Two Decades of Computer Based Instruction Projects: What Have We Learned?" *THE Journal,* Vol. 10, No. 2, 1983, pp. 88–95.

[15] Becker, L. A. and F. E. Petry, "Voice Input for Computer Aided Instruction Systems," *Journal of the American Voice Input/Output Society,* Vol 3, June, 1986, pp. 1–11.

7

Applications in Speech Synthesis

> Synthetic speech . . . not acceptable to critical ears . . . may well be acceptable if this means of communication creates opportunities that would not otherwise exist.
>
> A. T. Vincent[1]

Speech synthesis is the most accomplished computer speech technology. It is far more usable than speech recognition, speaker recognition, or language recognition, all of which have severe limitations that do not hinder speech synthesis. True, the comparison is somewhat unfair. Speech recognition of naturally spoken speech requires multiple levels of comprehension: phonetic, grammatical, and semantic—all requiring replication of human intelligence. Intelligible, if not natural, speech synthesis can be achieved without appeal to linguistic knowledge beyond the phonetic level. Also, we do not need to understand the process of language generation to do serviceable speech synthesis, whereas we need to understand the process of language understanding to do serviceable speech recognition. Furthermore, as noted in Chapter 5, we have not yet achieved a sufficiently deep understanding of the properties of the speech signal that are particular to the message, the speaker and the language, so that both speaker recognition and language recognition remain in an immature stage of development.

---

1. Quote is taken from [1].

7.1 "At the Tone, the Time Will Be . . ."

Because speech synthesis is a more mature technology, with historical roots in the 19th century, it is not surprising that applications of speech synthesis are more prolific, indeed, nearly endless. If you adapt the view of this book that even recorded human speech is to be considered "synthetic," we can trace the first application to the time-and-temperature service initiated by various telephone companies in 1931. The system consisted of phone lines interfaced to 78-rpm records containing over 4,000 utterances announcing the time at 10-second intervals, as well as advertising statements. In terms of sheer numbers this may be the most highly used application in history, having answered over six billion phone calls in its day.

For about 40 years the only commercially successful applications of speech synthesis used recorded speech. Although electronically synthesized speech had been invented in the late 1930s (see Chapter 4), it was difficult to implement and the voice quality was extremely poor. Two developments were needed to create the burgeoning demand for speech synthesis applications that we find today. One was the integrated circuit, which allowed electronic speech to become sufficiently intelligible. The second was text-to-speech, the computationally intensive activity of converting spelled text to phonetic text, and thence to actual speech.

Early text-to-speech applications required tolerance for the machine quality of the voice. It took another 20 years of progress in the design of computer hardware and software to make possible concatenative synthesis and allow the development of applications that require a more natural voice than purely electronic synthesis was capable of producing. (Of course electronic synthesis improved greatly during that period as well, but it still has a machine "accent.") Thus by the final decade of the 20th century, application areas of speech synthesis had proliferated to an extent once unimaginable.

7.2 When To Use Text-to-Speech; When To Use Digitally Recorded Speech

> Although a broad range of applications exists for text-to-speech technology products, the ones that will be most successful are those in which the vocabulary to be spoken is extremely large and changes frequently.
>
> Walt Tetschner, *Voice Processing, 2nd Edition*

All speech processing has tradeoffs. Speaker-independent speech recognition demands small vocabularies. Low impostor rates for speaker verification are traded for high false rejection rates for legitimate users. In speech synthesis the price of the naturalness of digitally recorded speech is a restriction to a small, fixed vocabulary; large, flexible-vocabulary applications must use text-to-speech together with either purely electronic synthesis, or a form of subword level concatenative synthesis, both of which lack the pleasantness and naturalness of human speech.

In particular, digitally recorded human speech concatenated at the word or phrase level is suitable when the texts are fixed, for example, as a greeting:

> Thank you for calling PhoneCare, Inc. If you are dialing from a touch tone phone, please press "one" now, otherwise please hold for an operator.

It is suitable when the text *structure* is fixed:

> The number is, area code {919, 783, 5528,} and AT&T will dial that number for a charge of {thirty} cents.

where the braces surround substitutable parts of the message. It is also suitable when it is possible to concatenate a small number of words and phrases to deliver all possibly needed messages.

> Your order will &not be shipped &until &month, &day, &year

where "&not" may be the word *not* or nothing; if it is the word *not,* then "&until" is the word *until;* if it is nothing, then "&until" is the word *on.* Clearly, "&month" may be any of the 12 month names; "&day" any of the 31 days of the month; and "&year" either this year or the next year. Thus output such as *Your order will not be shipped until January fifth, two thousand,* or, *Your order will be shipped on September eleventh, nineteen ninety nine,* are among the several hundred possible messages.

Text-to-speech is appropriate when the vocabulary is large, or unlimited. For example names and addresses, names of products, names of stocks, e-mail messages, fax messages, daily news reports, etc. It is also needed when the set of messages changes in unlimited or unpredictable ways, for example in reporting financial transactions, market prices, sports scores, etc. Text-to-speech is also generally preferable when messages are longer than one sentence, and when voice consistency is needed. Finally, text-to-speech can lower production costs since once installed, its use is essentially free, where as new digitally recorded messages require further human participation, editing, and other labor intensive activities.

The tradeoffs are summarized in Figure 7.1.

TEXT-TO-SPEECH	DIGITALLY RECORDED HUMAN SPEECH
More machine-like sounding	More natural sounding
Low storage requirements: only text need be stored	Large storage requirements (see Chapter 2)
Easily updated because text is easily updated	Time-consuming to edit and update
Consistent voice-quality and loudness	Problems of consistency
Intonation can be varied by rule	Intonation more difficult to control, especially in longer passages.
Most useful for large vocabularies, highly variable messages, long messages	Most useful when texts and text structure is fixed, and for short messages comprised of few individual vocabulary items

Figure 7.1 Text-to-speech versus digitally recorded human speech.

7.3 Interactive Voice Response Systems (IVRs)

One often hears systems that make use of synthetic voice described as interactive voice response (IVR) systems. The operative word here is *interactive*. If the listener has an opportunity to respond to the system and exert some control over what will be said next, the system is deemed interactive. Communication back to the system may be as simple as pressing keys on a telephone, or as complex as speech, for which a speech recognizer is used.

An example of a synthetic speech application that is not an IVR is the automated weather advisory reports available to pilots. The pilot listens, period. It is similar for the various warnings that one hears on shuttle buses, trains, elevators, and the like. The listener is totally passive and has no influence on the message, despite the joke going around that says that when the door closes on a passenger's face the system detects it and shouts "gotcha." There is a tendency among the uninitiated to call *all* synthetic voice systems IVRs and this, strictly speaking, is incorrect.

Most synthetic voice applications discussed in this chapter, and many in Chapter 6, are IVRs, and are referred to as such in the literature (but not in this book). For example the system that uses speech recognition to complete a collect or long-distance telephone call is an IVR because its various messages are in synthetic speech. Similarly, a bank interaction in which one indicates via touchtones what services are desired is also an IVR because the bank will respond to the various requests with speech synthesis. There is some debate as to whether a PC application that uses speech synthesis and speech recognition to teach a foreign language is an IVR. It is certainly interactive, but some insist that to be an IVR the communication must take place over telephone lines. Perhaps time will refine the definition more precisely.

7.4 Human Factors Revisited

Just as with speech recognition, human factors are the most important consideration in designing a system that uses speech synthesis, and it is difficult to imagine a system that couldn't be improved by attention to how the human interacts with the system.

In a book of this kind it is not possible to take up a study of human factors. College libraries are filled with books that do that. Rather, I would like to discuss two specific instances of where human factors influence system design.

Synthetic voices may be used as a warning alarm in the cockpit of military aircraft. Back in the 1980s the question arose as to whether the pilot would respond faster if the synthetic voice warning was preceded by an audible tone. The tone would serve as an "alerting cue." Experiments were carried out by human factors experts that determined that the tone actually increased response time. The alerting cue was amply provided by the unique sound of the electronic synthesis and the tone isn't needed [2,3].

In a different vein, Marics and Williges [4] carried out a classic human factors study concerned with how context and speech-rate affect the intelligibility of synthetic speech. They presented sentences of DECtalk® (electronic synthesis) speech at two different speech rates, 180 words per minute and 250 words per minute. Each of those cases was divided into three subtypes: one with no contextual clues; a second with a few contextual clues; and a third with heavy contextual clues, including context prompts during the experiment. Thus, there were six classes of sentences.

Intelligibility was measured based on transcription accuracy. Subjects listened to sentences in all six categories and were asked to transcribe what they heard. Errors were tallied, averaged, and presented in a table similar to Table 7.1.

All results were reported as statistically significant. That means that "some context" was truly better than "no context," that "heavy context" was truly better than "some context," and that speech rate truly affects intelligibility at all contextual levels.

Table 7.1
Average Error (%) for 30 Transcriptions by Contextual Cues and Speech Rate.

	Speech Rate = 180 wpm	Speech Rate = 250 wpm
No context	8.3	15.4
Some context	2.2	8.7
Heavy context	1.2	3.9

The significance of this human factors research is felt today, where designers of scripts for synthetic voice output carefully provide contextual clues as to the content of the message, and speech rates are inevitably within a few words per minute of, guess what: 180.

7.5 Application Areas

As with speech recognition, speech synthesis applications may be viewed on the sliding scale ranging from hypothetical applications to applications that are commercially profitable. The difference is that many more speech synthesis applications have been implemented because the technology is much less limited than with speech recognition. If you have a hypothetical speech synthesis application, technological limitations will usually not hold it back, although the general criteria of convenience, safety, health, entertainment, and economy still hold. The specific criteria of hands busy, eyes busy, remoteness and miniaturization are also present in implementing speech synthesis. In the 1980s, several auto makers installed synthetic voice in their upscale models to advise drivers of doors ajar, lights left on, low fluid levels, seat belts disengaged. This turned out to annoy drivers who, because their visual mode is usually not overloaded, did not need the voice output technology. They were willing and accustomed to receiving such information via a tone and by scanning the instrument panel. Today, automobiles no longer have this feature. On the other hand similar warnings in the cockpit are welcomed by pilots because their visual mode bears a far heavier load of information, and the use of the relatively free auditory channel is beneficial.

Speech synthesizers can speak at rates ranging between 100 and 400 words per minute. Optimal speaking rates are about 180 words per minute. However, the average person reads at a rate higher than the fastest intelligible rate of speaking, about 350 words per minute. Voice synthesis applications in which the computer speaks some kind of a report aloud is not suitable if the user is able to read the report. Thus an application to read e-mail over the telephone works because you have little other choice; but the same application to read e-mail while you are logged on to your system, permitting you to focus your eyes on another task will not work because it is natural and quicker to read e-mail, and there is little useful you could do with eyes free while listening to your e-mail.

Another failed application was "point-of-sale" speech synthesis that would announce the name of the product being purchased and its price as the customer checked out. A typical scenario popular a few years ago was in the "modern" supermarket. As the checkout cashier scanned the item, a synthetic voice would provide aural information. Major failure! The system succeeded only in annoying patrons and driving cashiers to distraction. Many quit to find other jobs. The problem, again, is that there was no need to be filled. A visual display of the item name and price suffices, and purchasers receive a detailed receipt, too. The fact that most customers were reading the tabloids as they were being checked out was to the contrary not withstanding. Indeed, it interfered with their reading up on Elvis's latest cameo appearance, engendering further resentment.

Sections 7.5.1–7.5.12 represent more of a sampling of application areas than an exhaustive listing. Most readers will know of areas I haven't covered and will often find themselves familiar with the application areas under discussion, an indication of how deeply synthetic voice has penetrated everyday life.

7.5.1 Aid for Persons With Disabilities

Speech synthesis has been a boon for the visually impaired, no doubt about it. Lesser known, but also significant, has been the use of speech synthesis for persons who have lost their speaking ability.

7.5.1.1 Aid for the Visually Impaired

No single technological advancement in history, with the possible exception of the invention of Braille, has had as positive effect on the visually impaired as speech synthesis. One of the earliest applications of speech synthesis was a reading machine for the blind, now manufactured by Kurzweil Computer Products, a division of Xerox Corporation. It combines an optical character reader that converts printed text into electronic form, and then uses a text-to-speech synthesizer to speak the text aloud.

The reading machines work very well on plain, unformatted text, which comprises most reading matter. It would be nice to see a more intelligent kind of reader capable of reading poetry in a metrically correct way, or able to read a play. The latter would require that scenery, stage and speaker directions, which usually occur amid the text of a play, be recognized and presented in a sensible way. Indeed, I can imagine—and imagining is what we're doing right now—a computer intelligent enough to switch voices when the characters change, at least to the extent of using a male synthetic voice for male characters and a female synthetic voice for female characters. Reading technical texts presents a crop of problems, including how to deal with equations in math books, chemical formulas in books on chemistry, and the special notations of physics and engineering. Tables, charts, and other kinds of figures are also problematic and are among the remaining challenges for an area that we might call "print-to-speech" technology.

Close your eyes and imagine interfacing with your computer that way. (Totally scary!) A text-to-speech interface for personal computers is nowadays standard equipment for blind or visually impaired students on the campuses of high schools and colleges throughout the country. Many adaptive text-to-speech programs are available for use in special disciplines. For example there are text-to-speech readers that can read lines of a computer language such as Fortran, allowing students to acquire programming skills and, indeed, to receive degrees in computer science and other disciplines.

There are "screen-to-speech" or screen reader programs under development that provide an aural description of the kinds of computer screens that are found on the World Wide Web, and in other computer applications. Such programs permit blind persons to interface to the Internet, download files, follow links to sites of interest, and so on. The validity of the Chinese proverb "one picture is worth a thousand words" tests the ingenuity of computer scientists who attempt to describe graphical interfaces in linguistic terms. Nonetheless, there has been a modicum of success, and each small step provides further assistance for ambitious persons trying to overcome the daunting handicap of a visual disorder.

The job of a screen reader is much more complex than simply finding text and reading it. It must be able to find and verbalize status lines, pop-up windows, dialogue boxes, error messages, icons, buttons, and other kinds of graphical interfaces. Currently there are a number of real, implemented, commercially successful screen readers on the market that help blind persons to use Microsoft Corporation's DOS and Windows operating systems, the IBM OS/2 operating system, the Unix X-Windows graphical user interface, as well as to assist users with software like dBase, WordPerfect, and Lotus 123.

Nearly every speech synthesis application will have a positive impact on the visually impaired, from talking vending machines to talking cameras. But one talking machine stands out as particularly apropos to the blind: the talking Everest Braille printer from TeleSensory Corporation. A Braille printer converts text to Braille so that a visually impaired person can have written records available. These may be class papers, general correspondence, or even something as mundane as a shopping list. The voice output of the printer assists the user in its operation.

Other speech output devices to aid the visually handicap are a paper money reader, which verbalizes the denominations of American currency, and a talking assistant that verbalizes the information contained in a dictionary and thesaurus. (This section uses some materials discussed in [5].)

7.5.1.2 Aid for the Speech-Impaired

Millions of Americans have a speech impairment. It may be as innocuous as a mild lisp, as serious as the distorted speech of persons with severe **dysarthria**—

disorders of articulation caused by nerve damage—or as crippling as total loss of speech. The causes may be congenital, the result of a disease or accident, or due to some kind of physical or psychological trauma.

If a person can interact with a computer in any little way at all, that person can have speech, however limited. The interaction may be no more than the ability to blow a puff of air or raise an eyebrow, which can be interpreted as a mouse click. Given this ability, the computer can display a menu of choices that can be scanned by clicking. Double clicking—two puffs of air—can run an application. An application permitting a severely disabled person to talk displays a screen full of the most useful words for communication, and an alphabet at the bottom of the screen. Through "clicks" the person can choose or spell a word, which is sent to a synthesizer for vocalization.

Dr. Stephen Hawking, one of the great physicists of this century, suffers from amyotrophic lateral sclerosis (ALS), known in America as "Lew Gehrig's disease." He is unable to speak. His movements are limited to the ability to raise an eyebrow. Nonetheless he wrote a best seller, *A Brief History of Time,* in which he explains the complex physics of the 4-dimensional space-time continuum. He used a sophisticated software package called *Equalizer,* which allowed him to construct sentences from a screen menu. To speak, and he is one of the most sought after speakers in the world, he uses a synthesizer to render his words vocally. Thanks to modern miniaturization, a small PC and a speech synthesizer can be mounted on Dr. Hawking's wheelchair, and he has speech wherever he goes.

A Hybrid System for Dysarthric Individuals

In the middle 1980s I had the privilege of doing research with a man with moderate dysarthria. He was originally a student in a class I taught, and when I first met him I didn't understand a word he said, which was awkward because he was not shy about asking questions in class. The other students helped me—some people are better at this sort of thing than others—and we got along for a few weeks until I realized that my hearing ability had adjusted to his speech and I could understand him. That being the case, I reasoned, there must be some consistency in the way he spoke, distorted as it sounded. The reason people who didn't know him couldn't understand him is that his speech deviated from the norm, and most people have high expectations as to what speech is supposed to sound like, and failure to meet those expectations is confusing.

A discrete-speech, speaker-dependent speech recognizer, it seemed to me at the time, would not have any such expectations. He could record words in his voice patterns such as they were to create training templates for a vocabulary. At a later time, when he spoke one of the words, the speech recognizer would know it, and could transmit the word to a speech synthesizer, which would articulate the word more intelligibly.

After a year of work we built a prototype system. The vocabulary was small, only 20 words, and the error rates we observed were between 10 and 20 percent, high by today's standards. Nonetheless it meant that our subject could communicate with the world outside his own circle, however slowly and awkwardly. Given the size of hardware in those days, the system had to reside in the laboratory as a curiosity item. The man could hardly walk around town with 200 pounds of electronic gear in a wheelbarrow. However, we had demonstrated, and reported to the National Institute of Health, the feasibility of such systems. (The results are available to the public in [6].)

7.5.2   Education

There are enormous opportunities for speech synthesis applications in the field of education. For the visually impaired, the advantages go without saying. Everyone, though, can benefit from interacting vocally with computers in areas ranging from vocabulary building, to learning a foreign language, to training oneself to correct speech defects.

7.5.2.1   Foreign Language Learning

Learning a foreign language is difficult for adults. It requires good analytical abilities, memory and phonetic skills. Repetitive exercises for ear training and speaking are necessary to achieve fluency in the spoken (as opposed to written) form of the language. The computer is the ideal task master for conducting both listening and speaking drills, which are tedious for a language instructor.

We saw in the previous chapter how speech recognition could, hypothetically, determine how accurately an utterance is pronounced, and thus provide feedback to the learner as to how good his pronunciation is. Speech synthesis can be used to provide the student with a model pronunciation to be imitated.

Speech synthesis can also be used to provide listening drills. The speech rate is easily adjusted to the student's comprehension level, as is the complexity and length of the utterances the student is asked to comprehend. Repetition is accomplished by pressing a key rather than rewinding a tape.

Actors and actresses often play a role in which they must speak "in dialect." The actress who plays Scarlett O'Hara in *Gone with the Wind* must deliver her lines in a southern accent, whether she is southern or not. A speech synthesizer can be programmed to speak in the target dialect and used as an automated speech instructor. Likewise, individuals who are raised speaking a minority dialect, and who wish to speak the standard dialect of a language, can avail themselves of computer programs that use a speech synthesizer to teach the desired dialect.

Language and dialect learning applications are commercially successful, and there is a wide range of products on the market for all skill levels, and for many of the languages in which American learners are most likely to be interested.

Translation of Spoken Speech

Either to aid language learning, or to aid the foreign traveler, there are computer programs that accept input in written form in one language, translate it into another language, and speak it. They do not do so well in translating the writings of James Joyce, but they work fine on "Where is the toilet?"

The simplest of these systems stores a set of stock phrases, similar to the ones that are found in travel guides, and speaks them in the foreign language. Usually such a system can also take a typed-in phrase in the foreign language and pronounce it correctly, an advantageous feature for persons who are good at reading and writing, but weak on pronunciation.

7.5.2.2   Speech Pathology

Most persons with minor speech defects can learn to improve or overcome them through repetitive practice under the supervision of a speech pathologist. Once the pathologist has explained to the patient what needs to be practiced, there is little reason for him, at $75+/hour, to administer the drills; a computer will suffice. The programming is very similar to that of foreign language learning. A synthetic voice speaks the target phrase, the patient repeats it, and using speech recognition, the computer determines how close to the target the patient came. Repetition is easy. Moving on to the next stage is easy. Making exercises more or less difficult to accommodate individual progress is easy.

The weak link in the system is the speech recognition, but even with the pathologist present to determine the accuracy of pronunciation, the computer can be a great help. It provides a consistent spoken target, its speed is adjustable, and it can keep track of the patient's progress at a level of detail impossible for the pathologist.

For parts of the country where there are few speech pathologists—perhaps one that passes through town once every two week—the computer is a way to get more frequent feedback. The patient goes through a set of drills and then transmits his speech electronically to a site where a speech pathologist can listen and determine whether the patient is improving and what steps to take next.

7.5.3   Emergency Scenarios

One of the earliest uses of synthetic speech was in the cockpit. It is used to reinforce emergency situations such as an imminent collision, low fuel levels, drop in oil or hydraulic pressure, the failure of a key component, an imminent stall (loss of lift), and so on.

Many locations where danger is always imminent, such as a nuclear power plant, have voice output systems for reinforcing visual and other kinds of warnings. Movie goers are familiar with scenes in which annihilation is imminent, and a very calm (usually female) voice is counting down in a tone as flat as a sheet of ice: "Fifty seconds to total destruction, and I couldn't care less because I am only a lifeless synthetic voice."

7.5.4 En Masse Advisories

When potentially catastrophic events are imminent it is often vital to inform a large number of persons whose job it is to ameliorate, and clean up after the disaster. Thus when earthquakes and hurricanes strike, or when a nuclear site has an imminent meltdown, the government and utility companies need to inform large numbers of workers where to go and what to do. This can be done automatically by telephone dial out systems that use speech synthesis.

Less vital but similar applications are found everywhere in society. Many school systems have an automatic dial-up system that calls a student's home when that student is absent. The systems are programmed to call repeatedly until they receive appropriate correct feedback. In schools where absenteeism is a problem, and parents are cooperative, the parent is given a special code unknown to the child, to be used to confirm reception of the information.

Catalogue companies and department stores are using synthetic speech to call up customers and inform them about the status of an order. Electric utility companies have automatic dial up systems that inform certain key customers, usually large consumers of power, that a brownout is imminent, requesting them to conserve energy until the critical period is over.

Another application was developed at North Carolina State University under my supervision. A computer was programmed to call up elderly and frail persons living alone and to check on their well being. A synthetic voice greeted them and asked them if they were okay. If they indicated they weren't, appropriate authorities were notified. If they didn't answer the phone after several tries, an emergency scenario was initiated so that someone could check on them. Several such systems were put "in the field," and are in operation today.

Finally, in another application that I had a hand in developing, a computer was programmed to dial up voters and, using synthesized speech, advocate a candidate in an election. A variety of different messages were used depending on where the voter lived, since location often determines the issues of highest interest. Thus the message could be personalized, to a certain extent. A follow up study was made after the election to evaluate the effectiveness of the telephone campaign. Results indicated that the system had influenced 6% of voters: 3% to vote against, and 3% to vote for, the candidate. Well, you can't win 'em all. The work is documented in [7].

7.5.5 Information Retrieval

Apart from aiding persons with visual disorders, the most natural application for speech synthesis is information retrieval. You're here, the computer with information you want is there, and there is a telephone line connecting you to the office where the computer is located. Help is just a phone call away, but wait a minute. What if the clerks are busy with other callers? What if the clerks misread information? What if the clerks are new and not familiar with the computer? What if the clerks are with their families at noon on the Fourth of July? No problem, because the clerks are no longer a barrier between you and your information. They have been replaced by an information retrieval system with a synthetic voice that is on duty 24 hours a day, never makes mistakes, and knows the system as intimately as a new mother knows her baby.

Many banks allow customers to call up and receive information about their accounts, and make certain transactions over the telephone. The bank communicates with the customer by synthetic voice. The customer most frequently uses the keypad, but speech recognition is not far behind. Similarly, one can often access a retirement account, receive information, reallocate resources, request withdrawals, adjust payment size, etc.

There are a number of services that allow you to dial up and request financial information of a public nature, such as the current price of various stocks, bonds, funds, commodities, options, and so on.

Transcontinental truck drivers have access to a service that they can call and receive a new or altered assignment from the company computer. Traveling sales personnel are able to call the home office computer and receive the latest sales figures, find out what is in stock and what has been back ordered, and have their e-mail read. Sports fans can get up-to-the minute sports scores and results. Travelers can receive the latest weather information. Many municipal airports provide a specific radio frequency over which a synthetic voice announces current weather conditions, including the all-important wind speed and direction, and the altimeter setting.

Another kind of information retrieval arises in the hands busy, eyes busy context of repairing or servicing a large machine such as an aircraft engine. Much of the information that is in a service manual resides nowadays in a computer. The technician may have cause to refer to the online manual during a repair or servicing task, especially if he or she is a novice. Often, however, the technician is engaged in some aspect of the repair operation and is unable to view the computer screen. If he is able to convey his question to the computer by some means—speech recognition would be ideal—the information needed can be returned by voice.

Similarly, in quality-control activities, the inspector may have need to query a data base as to what, say, a certain tolerance ought to be. A voice exchange would allow the work to continue without the major interruption of re-allocating hands and eyes to the task of handling the computer to retrieve the information.

Telephone directory assistance has long used digitally recorded synthetic speech. Other services such as a reverse directory, which provides the name and address that correspond to a phone number, use text-to-speech synthesis, speaking the information after it is retrieved. Synthesizers that specialize in pronouncing proper names are used for this task, as discussed in Chapter 4, Section 4.4.5.2.

7.5.6   Information Reporting

Information retrieval is receiving data from a computer in more or less raw form, with a few words of description. When you request a stock price you expect to hear something like: "PhoneCare Incorporated, high of four and five eighths, low of four and seven sixteenths, twenty three thousand, four hundred shares traded." Information reporting is a relatively new discipline in which a database of information is turned into a smooth flowing, natural language report by the computer, and delivered in either a written or spoken format.

If you are a Wall Street aficionado you probably read and hear a general report of stock market activities every trading day. It's not just the bare bones, but an interpretation of raw data into general trends. One hears expressions like ". . . technology issues were weak with IBM leading the trend, down one and one-half points to ninety one and a quarter. Only Intel managed to buck the downward slide; it was up one quarter to one hundred five and three-quarters. Trading volume was light." The idea is to take the raw data, make sense of it, and talk about it in complete sentences. This requires natural language generation.

Natural language generation is the opposite side of the coin from natural language understanding. The latter translates words into concepts; the former translates concepts in the form of databases into words. The field of natural language generation is a subarea of artificial intelligence, and as such is not taken up in this book. The literature is full of references which the interested reader can easily find.

The stock market report alluded to above is an actual implementation. Other implementations are on the way for sports reporting and weather reporting. Sports reporting would, for example, take raw baseball scores and produce the kind of report one expects to hear on a sports newscast. Instead of simply reading scores, sportswriters, and now computer/sportswriters, try to make it interesting. "In the clash of leaders in the American League East, the Boston Red Sox thrashed their arch rival New York Yankees nine to zip behind the five hit pitching of Mel Parnell. Dominic Dimaggio, Ted Williams and Bobby Doerr all homered for the Fenway Fans . . ." This as opposed to "Boston nine, New York zero," which is information retrieval. Weather reports would be generated in a similar manner, based on raw weather data.

An application many readers will be interested in, and possibly disappointed in, is astrological forecasting. A system, no longer in operation, would accept a caller's exact month, day and year of birth and read aloud that caller's "personalized horoscope." The system used a synthetic voice and left the impression that each caller did, indeed, receive personal information. It did so as follows. The system was equipped with the texts of 400 generic astrological "forecasts." The caller's date of birth was "hashed" into a number between 1 and 400, and the corresponding forecast was read by the text-to-speech synthesizer. The hashing function was designed so that a particular caller who called every day would receive a different forecast each time, leastwise for four hundred days. The set of forecasts would be refreshed from time to time. The application was successful. It didn't go out of business because it was exposed as fraudulent; it failed because the telephone company handling it went bankrupt.

A somewhat different kind of reporting is found in *Back Seat Driver,* developed in the MIT Media Lab [8]. This system would instruct a driver, in real time, how to find his way around the city of Boston, Massachusetts, using natural language spoken over a telephone by a speech synthesizer. In effect the system had to generate a set of coherent instructions based on a map of the city and the driver's location, and then render those instructions into clearly phrased natural language suitable for directing the driver of a vehicle. The system had to have a good sense of timing, and issue its instructions piecemeal as the driver approached the target—it wouldn't be useful to give all the directions at the outset. In addition to directional information, Back Seat Driver was clever enough to offer advice and warnings to make the drive safer. For example, it may say "Get into the left-hand lane because you are going to take a left at the next set of lights."

Applications for which Back Seat Driver is a prototype have been developed by automobile rental companies. For example, one company has a product called NeverLost®. This is an excerpt of what the company says about it on their Internet site. (http://www.hertz.com/serv/us/prod_lost.html#top)

. . . NeverLost calculates the route and issues visual and verbal commands to guide customers to their destination.

7.5.7 Electronic Mail and Fax Readers

Historians will have plenty to say about the 20th century. Among it all, they will identify this as the century of addictions. Apart from the usual suspects—tobacco, whisky, and dope—one finds sexual addictions, shopping addictions, gambling addictions, addictions to sports, addictions to television, addictions to exercise, and one of the latest: addictions to e-mail. Yes, we're told, some people have to have their e-mail fix every day. They become nervous when they travel away from their computer; trips are postponed, visits are cut short, affianced ones are left standing at the altar, all to feed the insatiable appetite of the e-mail dragon.

What technology createth, let technology satisfieth. It is now possible to access e-mail by telephone. A system interfaces to your mail box, scrubs the mail clean, extracts critical information such as who the sender is, and reads the text of the message to you. Eventually, no doubt, when speech recognition catches up, one will be able to compose verbal answers and have them sent as return e-mail. In any case the e-mail craving may be sated by a telephone call.

Scrubbing is the removal of irrelevant material that accompanies the text of an e-mail message, leaving behind the actual message. It is effortlessly accomplished by the sighted human viewing a screen, but imagine the difficulties of extracting the relevant information from the typical e-mail message shown in Figure 7.2. It was sent to me precisely as shown, in answer to my request for the name of a new Japanese restaurant in Raleigh.

Suppose this e-mail message was accessed by telephone and was to be read by a text-to-speech synthesizer. Apart from the message itself, what other information is pertinent? Certainly the date and time of the message, the sender and probably the subject. This information is available. The date and time are given twice, in two different formats; the sender is identified at the top of the message by his e-mail address, which would work in this case since I know that address but might not work in general. The sender also has a signature at the bottom that might be retrieved if the computer knew that it appeared immediately under the double hyphen (the ones not preceded by >). The subject is "question mark," and comes after "Re:" and that, at least, is easy to locate and read. Its lack of informativeness is my fault, not the system's. Otherwise, the first seven lines of the message should be ignored. A text-to-speech reader could read those lines; it can read *anything,* spelling out "words" it cannot decipher and naming symbols by saying "dollar sign" or "right-angled bracket" but that is

Message 4/5 From tkl@csc.ncsu.edu   Jun 25, 98 01:01:26 pm -0400

Date: Thu, 25 Jun 1998 13:01:26 -0400
To: Dr Robert D Rodman <rodman@adm.csc.ncsu.edu>
Subject: Re: ?
X-Phone: 919-515-5343
X-Fax: 919-515-7896
X-WWW: http://www.csc.ncsu.edu/~tkl

As written in the ancient tomes by Dr Robert D Rodman (rodman@adm.csc.ncsu.edu):
> Tim,
>
> Tell me again what the name of the new Japanese restaurant is, Kuki's
> replacement. I'm going to make Zack take me there for Father's Day.
>
> -r
>
> --

Waraji: Next to the public library on Duraleigh Road (slightly before the
Glenwood Ave and Duraleigh intersection). Next to the big metal statue in
the shopping center.

--
Tim Lowman: Sys Prog II | SMail: NCSU, Box 8206 Raleigh NC
Email:tkl at csc dot ncsu dot edu | Postmaster at csc dot ncsu dot edu
All language designers are arrogant. Goes with the territory... :-) --
Larry Wall

Figure 7.2 A typical e-mail message.

undesirable in most instances when reading e-mail because the information is
not relevant.

Line 8 is the next line of interest. It indicates that the message is a reply
to a previous message. So far, so good, but one hopes that the e-mail reader
will omit reading my own e-mail address back to me. The next eight lines, the
ones preceded by >, indicate the previous message, and should be read, ignoring the >. The actual message might not make sense if I had forgotten
what it was in response to. In this particular case, reading the repeated material is probably okay. However, many e-mail messages are answers to answers to
answers to answers to questions, and those of you who use e-mail know I am
not exaggerating. Should all previous exchanges be read, or can the system figure out a way of giving you the option as to whether to read the various, previous messages?

At last we come to the message itself, beginning with "Waraji ..." This
can be read by the text-to-speech processor. After reading the message we would

expect the e-mail reader to give us the option of hearing the message again, composing (somehow) a reply, or moving on to the next message. The final four lines of the message should be scrubbed away, once the name of the sender has been extracted.

This is not a contrived example. It is an actual e-mail sent to me today before I even decided to put an actual e-mail message in this book. By the way the "signature" at the bottom, given as "tkl at csc dot ncsu dot edu" rather than as "tkl@csc.ncsu.edu" is designed to foil mass marketers who swipe e-mail addresses and put them in e-mailing lists. Their program doesn't recognize the verbal, scrubbed form as being a valid address, leastwise not yet.

Fax readers face many of the same kinds of problems that e-mail readers face in needing to decide what information should be skipped and how to extract the relevant information such as who the sender is, what the return fax number is, and so on. Additionally, the fax has to be rendered in a text form suitable for a text-to-speech synthesizer.

What we may deduce from the previous several paragraphs is that consistency is needed among the various components of an e-mail system: the sender, the receiver, and now the reader. The sender is concerned with formatting outgoing messages; the receiver is concerned with displaying incoming messages. In general, two different e-mail systems can interpret each other's e-mail because it is mostly a matter of screen display—the human reader does the scrubbing. The e-mail reader changes this balance. It needs to have some idea in advance as to the formatting of incoming messages if it is to have a fair chance to scrub the message clean of extraneous information. What is required here is a repeat of similar scenarios that have occurred and re-occurred since the invention of the computer, namely *standards*. If e-mail messages are formatted according to some set of international standards, then an e-mail reader could take advantage of that information when preparing a message for speech.

7.5.8 In the Dark

Two scenarios have been discussed in the applications literature where voice is needed because of the need to maintain perfect darkness, making reading impossible. One is military black-out conditions. A military outpost in a combat zone may have the need to maintain darkness to prevent detection. At the same time, interaction with a computer may be desirable. Voice, in both directions, is the answer.

The photographic dark room is another environment where total darkness is required, but where it is often desirable to obtain information from a computer. Speech synthesis can serve that need.

7.5.9 Toys and games

It is a toss-up as to what application area of speech synthesis was the first to achieve wide success: aids to the visually impaired or usage in toys and games. As is always the case with technology, it can be used equally well for good, evil, and trivial. One could argue that its use in toys and games is shallow compared with its nobler usages, but then our children need to play (and so do we), and speech synthesis makes that play more interesting, and in one outstanding case, more educational.

7.5.9.1 Speak 'n Spell™

An early application of speech synthesis, using LPC compressed speech on a single chip, was a child's educational toy called Speak 'n Spell™ by Texas Instruments. It had buttons of letters which a child could press and hear the name of the letter pronounced. It also had various spelling games that the child could play designed to teach her or him the alphabet and rudiments of English spelling. One feature was the ability to play hangman, a word guessing game in which one tries uncover the hidden word by guessing what letters are in it. I would be lying if I didn't say I played the game for many hours with my own children while they were growing up and always found it a pleasure. Surely this device will go down as one of the 10 best toys of our century.

Numerous other such gadgets are available at your local mall. Most of them are clever, well-designed, educational applications of speech synthesis that outweigh the lamentable use of the technology to provide the grunts, groans, cries, and shouts of violent arcade games.

7.5.10 Transportation

I already mentioned, under information retrieval, how truck drivers can received information from the company computer over the telephone via synthesized voice. In fact, the entire transportation industry is using voice output to increase efficiency and reduce costs.

In the trucking industry it is possible to receive automated information by voice regarding rates of freight movement, location and status of shipments, the status of permits, and the estimated times of arrival of particular drivers. Drivers may also receive both official and personal messages via phone.

In the airline industry, as most of us know, flight information is available by telephone to the public; crew scheduling may be accessed by crew members with the need to know. Similar kinds of information are dispensed via speech synthesis in the rail industry.

In hotels, guest messaging and wake-up service calls are implemented using speech synthesis.

7.5.11 Government Services

Government offices receive many calls requesting information. These range from tax information from the Internal Revenue Service to the time and place of town meetings. Much of this information can be dispensed by use of speech synthesis over phone lines. To mention a few applications, we have jury duty-related information, tax information, road closing information, lottery results, school lunch menus, opening and closing times of public buildings, and un-employment claims processing.

7.5.12 Disguise

As noted earlier, technology is amoral and can be used for evil. If a person wants to leave an anonymous phone message that cannot be traced to him through his voice, the perfect method is to use a popular text-to-speech system and let it do the talking. A given make of synthesizer sounds identical to any of its brethren and provides the perfect shield for unscrupulous persons who wish to leave blackmail information, bomb threats, and the like.

References

[1] Vincent, A. T., "Computer Assisted Support for Blind Students—The Use of a Microcomputer Linked Voice Synthesizer," *Computers and Education,* Vol. 6, 1982, pp. 55–60.

[2] Simpson, C. A., and D. H. Williams, "Response Time Effects of Alerting Tone and Semantic Context for Synthesized Voice Cockpit Warnings," *Human Factors,* Vol. 22, No. 3, 1980, pp. 319–330.

[3] Hakkinen, M. T., and B. H. Williges, "Synthesized Warning Messages: Effects of an Alerting Cue in Single- and Multiple-Function Voice Synthesis Systems," *Human Factors,* Vol. 26, No. 2, 1984, pp. 185–195.

[4] Marics, M. A., and B. H. Williges, "The Intelligibility of Synthesized Speech in Data Inquiry Systems," *Human Factors,* Vol. 30, No. 6, 1988, pp. 719–732.

[5] Lazzaro, J. J., "Talking Personal Computers," *Journal of the American Voice Input/Output Society,* Vol. 13, March 1993, pp. 17–26.

[6] Rodman, R. D., "Computer Speech Recognition in Augmentative Communication," *Journal of the American Voice Input/Output Society,* Vol. 6, July 1989, pp. 61–76.

[7] Rodman, R. D., and J. O. Williams, "The Use of Computers in Election Campaigns," *International Journal of Speech Technology,* Vol 1, No. 1, 1996, pp. 33–40.

[8] Davis, J. R., and C. Schmandt, "The Back Seat Driver: Real Time Spoken Driving Instructions," *Vehicle Navigation and Information Systems,* 1989, pp. 146–150.

8

Applications in Speaker Recognition, Language Identification, and Lip Synchronization

> Language was born in the courting days of mankind; the first utterances of speech I fancy to myself like something between the nightly love lyrics of puss upon the tiles and the melodious love songs of the nightingale.
>
> Otto Jespersen, *Language, Its Nature, Development and Origin*

Speech recognition and speech synthesis are attempts to replicate a biological process at which humans excel. Language is as intimately a part of being human as opposable thumbs and bipedal gait, and its role is indispensable in the evolutionary processes that culminated in *Homo sapiens*. Human performance defines the height of the bar over which speech recognition and synthesis must leap. Conversely, humans can hardly be said to excel at speaker recognition or language identification; these processes are not innate to the human species. The difference between speech recognition and synthesis on the one hand, and speaker recognition and language identification on the other, is comparable to the difference between oral language and written language. Oral language is an organic part of being human, acquired naturally by children without deliberate instruction, independent of the level of intelligence;[1] reading and writing skills

---

1. This notion, which some readers may find controversial, was expressed by Descartes in "Part V" of his *Discourse on Method:* "It is a very remarkable fact that there are none so depraved and stupid, without even excepting idiots, that they cannot arrange different words together, forming of them a statement by which they made known their thoughts."

are learned, if at all, by dint of deliberate effort, made easier by superior intelligence. (These views are also articulated in [1].

The motivation for applications in speech recognition and synthesis must be substantial since machine performance is expected to be inferior to human performance. In most cases we'd use a human if it were practical. The motivation for applications in speaker recognition, language identification, and lip synchronization is that machine performance is superior, in addition to other criteria for automation.

8.1 Applications in Speaker Recognition

Speaker recognition applications are far fewer than applications in either speech recognition or speech synthesis, but they are growing the most rapidly. According to [2], revenues from speech recognition in millions of dollars are 970 in 1998 and will be 2,093 in 1999; from speech synthesis, 131 in 1998 and 231 in 1999; and from speaker verification 27 in 1998 and 77 in 1999. The growth rates are 216%, 176%, and 285%, respectively. To imagine the application potential, think about how many times a day you use a key, a card with a magnetic strip, a PIN number or a password. Hypothetically all those actions could be replaced by your voice.

I have divided speaker recognition applications into six broad areas. As in Chapters 6 and 7, these divisions have amorphous boundaries, with some applications belonging somewhat to one, somewhat to another. The field of speaker recognition is wide open and dynamic, and new applications are continually being implemented.

Speaker recognition applications may be categorized as *access, authentication, monitoring, fraud prevention, forensics,* and *personal services.* These are described in Sections 8.1.1–8.1.6.

8.1.1 Access

Access refers to the physical entry of a person into a secured area, or the electronic entry into a secured computer or computer file. Access can be restricted to persons whose identity is established by speaker verification. Such voiced-based authorization is often a part of a security system that also includes the use of PIN numbers, passwords, and other more conventional means. That is because speaker verification is not a fully perfected technology.

I probably work at one of the few remaining "business" sites that does not require employees to show identification to gain access to the workplace, namely a large university. I suspect that will change before I retire. Only a month ago (May 19, 1998) one of my colleagues, a diminutive, partially disabled middle-aged woman was brutally beaten in the halls of a university building in the center of campus during daylight hours. She spent two weeks in the intensive care unit, and she will never be the same. Hardly a month goes by when a crime of violence doesn't take place on the campus—muggings, rapes, shootings—and the months that violence misses, larceny fills. Similar stories are to be found on many other campuses.

Corporations no longer allow free access to their buildings. Employees and contractors, at the very least, are required to display badges, and in sensitive areas further means of access control such as retinal scans or speaker verification may be imposed. Furthermore, most corporate sites have layers of secure areas. Every employee may be entitled to enter the outer layer, but only certain employees may have access to inner layers. Speaker verification, which identifies an employee uniquely, may be used to enforce the various levels of access control. It may also be used to monitor the time a person spends in the workplace by the simple expedient of controlling both entering and exiting. Use of voice would make this simple and unobtrusive. It's Orwellian, but there you go.

Home access, too, is subject to layers of security. All family members need access to the outer doors, but only certain members may be allowed access to the computer, certain cable channels, the security system, or the safe containing the family jewels.

To date, speaker recognition systems have not penetrated the home security market very deeply, though one security company is testing a system that can be disarmed by a voice password spoken by a designated person.

Car access is yet another area where voice-based security systems are gaining ground. One of the largest automobile manufacturers is testing a speaker identification system to control door locks and ignition switches. One interesting twist to this application is that the ignition switch can be programmed not to work if the driver is under the influence of drugs or alcohol, since intoxication is detectable in the speech signal [3].

Even country access is benefiting from progress in speaker verification. The *U.S. Immigration and Naturalization Service* (INS) has installed a speaker verification system at the port in Scobey, Montana, between the United States and Canada. Residents in that area need to travel between the two countries on a daily basis—farmers actually live in one country and have fields in the other. Rather than keep personnel on duty around the clock for the rare nonresident travelers, a gate controlled by a verification system restricts access but permits residents who have registered to open the gate by speaking their password. Nonregistered persons may only pass through the gate during daylight hours when an agent is on duty. There are several hundred people enrolled in the system, which is used about 40 times a month.

Electronic access is not a new idea. The use of passwords to control computer and file access has been around since the first computers. Telephone and network communications have vastly expanded the opportunities for electronic access and the need to restrict access to legitimate users has grown concomitantly. Nowhere is this need more vital than in the banking industry.

Millions of banking and credit card transactions occur daily. Account access is generally protected by the use of account numbers, PIN numbers, and passwords, including the ubiquitous "mother's maiden name." All these can be stolen, and many banks are looking at biometrics as a way to ensure that a person is who he says he is. Since most transactions take place over phone lines, speaker recognition is the biometric of choice.

A small independent bank outside of Chicago provides a typical case study. In the words of one officer, the bank wanted to "provide additional assurance that it is the right person who is doing the transfer." The bank ran a pilot study using its employees to determine the feasibility of speaker recognition. This proved to be successful so the bank next informed customers of the opportunity to achieve additional security when making phone transactions. The bank conducted an extensive promotional campaign that included personal letters, mass mailings, brochures available at the bank, and personal contact. It deployed its system in mid 1996.

Customers were enrolled, with their consent, the first time they called the system. Since the system is text-dependent speaker verification, customers enrolled by repeating their social security number five times to complete the training phase and establish the speaker models. Customers who had trouble doing this were urged to call the bank for help. The fact that the bank's employees had already gone through the process was of immense use because many of the difficulties had been experienced, and the employees were well-seasoned and better able to aid customers.

The bank requires customers to use voice verification in combination with a PIN number if they want to use the automated system to transfer funds between accounts. Nearly every customer who wants to use the system is able to do so, and most appreciate the additional security that the voice verification provides.

One lesson derived from this brief case study is the importance of human factors. At every stage of deployment, from inception to enrollment, the bank paid attention to the needs and concerns of its customers. They were rewarded in the end by a workable system and customer satisfaction. The bank is planning to offer further services using both speech and speaker recognition.

Other applications of speaker recognition for electronic access control are securities trading, telephone toll calls, retirement account information, and access to the Internet. Laboratories, computer rooms, filing cabinets, machine-equipment control, and elevator floor control are physical sites whose access may be limited by voice.

8.1.2 Authentication

Authentication is similar to access restriction in that speaker verification is used to certify the identity of a person. Authentication is the term I have chosen for certification for reasons other than access.

Did you ever cut school when you were a kid (assuming you're not, now)? Cutting was made difficult by the fact that the school would call your home to ascertain the legitimacy of your absence. If you cut on a day when your parents weren't home, and you were a good mimic, you could take the call yourself and assure the school counselor that you were sick and dying and would never be heard from again. You might fool the counselor, but you wouldn't fool a speaker verification system. In some school districts the adult responsible for the student registers his or her voice with the school. When a call regarding absence is placed, the person to whom the absence is reported must be authenticated by speaker verification, otherwise the school takes further action.

Speaker verification may be used to authenticate borrowers at public libraries, rental store customers, persons receiving licenses, and data or financial couriers. Several states are considering collecting voice samples when issuing drivers' licenses and authenticating persons at the time of renewal or when replacing a lost license. This is easily done by having the candidate read the unique driver's license number several times as part of the issuance protocol. A further advantage of collecting voice samples is their potential use in forensic situations, discussed in Section 8.1.5.

8.1.3 Monitoring

Overcrowded prisons are among the top social problems facing the United States as it approaches the new millennium. A few days ago the *Raleigh New and Observer* newspaper[2] ran a photograph on page one of the interior of the Wake County jail. Not only were the cells packed to capacity, but several hundred inmates were "housed" in the hallways, where each had but a few square feet of floor space containing only a mattress. The scene is not an uncommon one in the nation's jails.

---

2. *The News and Observer*, Sunday, June 28, 1998, p. 1A.

Many persons who are incarcerated in a jail would do well under some form of house arrest, thus relieving the overcrowded jails. They would, of course, require monitoring, and that is where speaker verification comes in. Randomly scheduled phone calls are made to the locus of arrest. The convict must take the call, speak the required phrases, and be verified. Failure to do so results in loss of the house arrest privilege.

In general, such a system can be used for the electric monitoring of offenders on probation, parole, work release, and pretrial release. Since the person to be monitored is not necessarily at a certain phone number, the system uses the page/call-back method. Random calls are made to the offender's pager, who must then call an 800 number within 10 minutes and be verified. The system has the additional advantage of using caller ID to determine from where the call is placed. This can be used to enforce the restriction of movements sometimes placed on certain types of offenders such as perpetrators of domestic violence or purveyors of drugs. Offenders are motivated to make the system work because they prefer not to wait in jail and because it can mean a reduction in bail.

A further advantage of such monitoring systems is they can be used to communicate with offenders, such as reminding them of trial dates and other appointments or advising them of a change in their status. Moreover, a report of activities for each offender may be generated for future use. This is a real, implemented, commercially successful application. Some offenders object to the 10-minute restriction as an imposition, but in all other respects the system performs adequately.

A company called T-Netix—check out their Web site at http://www.T-NETIX.com—features a product called *Inmate Calling Service.* It uses speaker verification as part of a system designed to control telephone calls that prisoners are allowed to make. Different prisoners have different calling rights, so to prevent a prisoner with restricted calling rights from impersonating a prisoner with unrestricted calling rights, each caller's voice is monitored and subjected to verification.

Though not currently a part of Inmate Calling Service, the following expansion of the service is plausible. Many prisoners are permitted to call only their attorney, a spouse, a child, or a parent. Those persons' voices may be registered in advance, and later verified before a call is connected, and while the call is taking place, thus ensuring the convict does not talk with a forbidden person.

A different kind of monitoring system is currently in the research stage of development. It is a computer that calls elderly or frail persons living alone and determines from their voice whether their health is deteriorating to a point where help is needed. This technique would supplement the monitoring service for the elderly described in Chapter 7, Section 7.5.4.

8.1.4 Fraud Prevention

Many of the applications already discussed could be interpreted as having the purpose of fraud prevention. This section mentions applications whose specific purpose is to deter fraud *in a business where fraud is a problem.*

If you haven't been a victim of cellular phone fraud, I'll bet you know someone who has. My cell phone was "cloned" last year and used to make international calls from Philadelphia, and I know several others who have had similar experiences within my own small circle. It is a multibillion dollar rip-off and an embarrassment to the wireless carriers, which are reluctant to talk about the scope of this form of larceny.

Never fear, though, help is on the way in the form of speaker recognition. The technology still requires improvement so that it works reliably over channels that are partly, if not entirely, wireless, but that age is nearly upon us. A typical use of speaker verification fraud deterrent would apply to a cellular phone that was used primarily by one person. That person would register his or her voice with the carrier. When a call is placed from that number, one of two verification processes may occur. A computer may interrupt the call and ask the caller to say a password and use that utterance for verification. This, of course, is intrusive and annoying. A second, better choice is for a computer to listen in and do text-independent verification. A verification failure may be followed by an interruption and a text-dependent verification. There are many ways to structure the anti-fraud protocol, some of them dependent on how highly the customer is willing to be inconvenienced to prevent fraud.

AI may be applied. Cellular phone fraud often follows a pattern that ordinary usage doesn't, namely multiple calls to a foreign country within a short period of time. This could be detected and used to trigger a verification interruption. More generally, a person's calling patterns could be modeled from past usage, and any deviation from those patterns would trigger a verification procedure.

Only text-dependent speaker verification has been deployed commercially to prevent phone fraud, and at this time with little penetration. The interested reader will find further information at the URL: http://www.t-netix.com/Markets/Wireless.html.

Another application of fraud prevention is at the University of Maryland at College Park. One of the services the university provides is *Voice Access to Voice Mail* (VAVooM), an 800 number for faculty and staff to access their voice-mail boxes. To prevent hackers from using VAVooM as a gateway to toll fraud, the university uses speaker verification. Only university members who enroll in the system by providing voice templates are allowed usage. As of the middle of 1997 about 250 faculty and staff availed themselves of the system. The telecommunications department at the university reports no unauthorized entries since the system went into effect.

The system is text-dependent. Users are required to repeat their password up to 15 times in order to establish their speaker model. False rejection rates are relatively high because the university has adjusted the system's threshold to require a very close match between the training and operational exemplars of the password to thwart impostors. (See Figure 5.2 and surrounding discussion.) Calls over poor channels such as from pay telephones often fail, and the user may be requested to re-enroll over that telephone if it is to be used often. Before doing so, of course, the caller must verify his or her identity in some other way.

This system typifies the state of the art in speaker verification at this time: text-dependent with a somewhat inconvenient enrolling procedure and a barely tolerable false rejection rate if impostors are to be kept securely at bay.

In Chapter 6 I mentioned the use of speech recognition as an aid to handicapped voters. Speaker recognition could be used to prevent voter fraud, and ultimately, to permit voting by telephone. This is a hypothetical application that would require a national effort and a large expenditure, but it may be an effective way to engage more voters in the electoral process. The application, to be secure, would be one with both text-independent speaker verification and speech recognition. The voter would dial up and be prompted by the system to repeat a randomly chosen phrase. That the correct phrase was repeated would be verified by speech recognition and that the correct person spoke the phrase would be verified by speaker recognition. This procedure could also take place at a polling station.

8.1.5 Forensics

We saw in Chapter 5 that the history of speaker recognition is closely tied to forensic activities. The FBI, military security, and state and local law enforcement agencies have all been concerned with identifying and verifying individuals based on their voice.

There are two types of application in forensics: showing that two exemplars of voices were made by the same person or not, and showing that a voice exemplar was made by a particular person from among a set of persons. There are three levels of confidence of significance: confident enough to be evidence in court, confident enough to influence the investigation, and not sufficiently confident to affect the investigation.

Law enforcement agencies sometimes conduct a "voice lineup," in which witnesses listen to several voices and attempt to pinpoint the one used in a crime. A voice lineup may also be conducted by computer in cases where there is a recording of the criminal's voice and a group of likely suspects. In such a case models are built from the voices of each suspect and compared with a model based on the exemplar associated with the crime. Current technology is capable of the second level of confidence, which can help the authorities narrow their search for other evidence by eliminating some suspects and pinpointing others.

A common scenario is to be in possession of a recording associated with a crime, and to have a suspect. The goal is to come up with some kind of likelihood that the suspect made the recording. Given the legal principle of innocent until proven guilty, this kind of forensic verification must take into account the vast number of other persons who may have made the recording. Even a match that is 99.99% confident allows for 100 other persons in a population base of one million.

One approach that might be taken is to find **cohorts** of the suspect. A cohort is a person whose voice is similar to that of the suspect, usually a person of the same sex, of similar size and age, and speaking the same dialect. Models of all the cohorts and the suspect are made and compared with a model of the recorded voice. If the suspect's model is significantly closer to the recorded model than the models of all cohorts are, there is excellent reason to pursue the case, even with as few as 10 cohorts, if the cohorts are carefully chosen. On the other hand, if the suspect's voice model is the same distance from the recorded model as several of the cohorts, one might consider the suspect exonerated.

I suggested earlier the possibility of a national database of adult voices, collected through driver's license bureaus, the social security agency, or as part of voter registration. There may be some legal problems here, but if the Supreme Court will let your employer force you to urinate in a cup for purposes of drug testing, I suspect that there will be little blanching at the prospect of a national database of voices.

There is a need. The 1996 Olympic Park bombing, in which a warning from someone (the bomber?) was recorded, may well have been solved had such a database been available. However, let me demonstrate with a mundane example garnered from today's newspaper.[3] Here is an excerpt:

---

3. *The News and Observer,* Tuesday, June 30, 1998, p. 3B.

Two hours after the shooting, a man mysteriously called 911 from a pay phone at a North Raleigh shopping center to report that someone had been killed by the tracks downtown. In May, police asked area TV stations to play a copy of the 911 tape in hopes someone would recognize the voice.

Suppose we had a national database of voices. The voice on the 911 tape could be compared with voices in the database and results would surely be as useful as an appeal to a TV audience.

Although it seems like a vast amount of data, it isn't. An excellent voice model can be achieved using 500 bytes. (It is claimed in [4] that it can be done in 40 bytes.) Assuming 100 million voice files, we have 50 billion bytes of data, or 50 gigabytes, which would fit very nicely on two or three hard-drives that you could carry around in the glove compartment of your car.

Finally, imagine the role the national database would play in crime control. The telephone is used in the course of committing much crime, and legal wiretaps abound. Now the identity of the talkers, in many cases, could be ascertained. Just knowing of the database, and that their voice is in it, would deter many swindlers from approaching their targets by phone for fear of being recorded and identified.

Many readers will be appalled at my suggestion, which admittedly smacks of Big Brother in George Orwell's *1984*. However, 1984 has come and gone, and in this, the computer age, there is already a tremendous amount of information available to the public and to the government about nearly everybody. For example the North Carolina State Department of Motor Vehicles had plans to sell a database that included all the information ordinarily found on a driver's license *including the picture* until public outrage forced them to cease. There is no closing this floodgate. Our only protection is to be found in our legal system, and that would be the individual's shield against an abuse of the national voice database, whose sole use would be the deterrence and punishment of crime.

8.1.6 Personal Services

One can imagine a speaker recognition system in your automobile that serves as both door and ignition key and automatically adjusts seats, mirrors, windows, and the sound system to your personal liking. This is especially useful when two or more people share one car.

One can imagine a speaker recognition driven system for rationing candy to children, medication to forgetful adults, donuts to persons both dieting and not dieting, and so on.

Finally, one application that is real and will soon be deployed is a voice-directed video camera for video conferencing. The system recognizes the voice of the speaker, knows where the speaker is located, and transmits the appropriate image.

8.2 Applications in Language Identification

Currently there are no automatic language identification systems in commercial use. Expect this to change soon. There are many in various stages of development in various research laboratories around the country, such as Bell Labs, Texas Instruments, and the Oregon Graduate Institute.

The technology is advancing rapidly and will soon be ripe for implementation. Sections 8.2.1–8.2.3 describe some of the potential application areas.

8.2.1 Telecommunications

International telephone call routing is likely to be the first commercially successful application. Based on the caller's speech, the language is identified and automatically routed to an operator or interpreter trained in the language the caller is using.

In the more distant future, when automatic machine translation has reached a serviceable level, the language identification system will function as a front-end. The caller's language will be detected and the call routed through a machine interpreter that accepts the caller's language as the source, and the language normally spoken in the location being called as the target. This may not work all of the time as people may not speak the local language. Presumably—this is all hypothetical—the caller can indicate the target language in special cases.

8.2.2 Communications Monitoring

The U.S. National Security Agency has a deep and abiding interest in automatic language identification and has invested money in research in the area. One can only guess at the reasons for its fascination. Surely one is that the agency monitors the international air waves—radio, television, and other wireless communications. Much of what agents overhear is in foreign languages. Automatic language identification would assist in routing the information to the proper channels for interpretation.

No doubt about it, industrial spying happens, and no doubt too, wireless communications are monitored as part of that spying. Since many large corporations are multinational, their employees communicate in a variety of languages. The spying entity would use automatic language identification to direct the overheard conversations to the appropriate translating authority.

8.2.3 Public Information Systems

Public service agencies, and emergency services in particular, must be prepared to assist persons speaking a foreign language. Often there is a pool of interpreters available who speak a variety of languages, one of whom could help, but how can the person who answers the phone choose the right one? An automatic language identification system can solve that problem, providing its database includes the unknown language.

8.3 Applications in Automatic Lip Synching

The commercially successful applications in automatic lip synching all require that the speech be computer generated, or based on a preknown text. Automatic lip synching of spontaneously spoken speech is still in the research stage, but several interesting hypothetical applications are already being considered.

8.3.1 Animation

By far, the most successful use of automated lip synching is found in the animation/film industry. Most animation you see today is not lip synched. It is too labor-intensive to do so. The mouth of talking characters moves, but it does not move in such a way as to replicate a mouth articulating the speech being spoken. Today's audiences demand a higher degree of reality than ever before. High-quality animation such as that seen in the motion picture *Toy Story* requires realistic lip synching. While this job can be done by hand, frame by frame, it is better done with the aid of a computer. Several companies market software that, to a greater or lesser degree, automates lip synching, including Industrial Light & Magic Commercial Productions, Autodesk, Inc, and Alias Wavefront, a subdivision of Silicon Graphics Company. Among the various such products available is Autodesk's 3D Studio Max.

Use of automated lip synching is desirable wherever high-quality cartooning is needed, in particular, whenever there is a talking head to be animated. In a typical application a human voice is used for the cartoon character's voice, although computer-generated voices are becoming more common. In both cases, the text of the talk is known, a phonetic transcription is created automatically from the text, and the mouth shapes are chosen to correspond to the sounds being pronounced. The timing of the mouth movements is done solely by the animator in the case of overlayed human voice since none of the available software is sophisticated enough to do it automatically. Computer-generated voices can provide timing information to make the animator's job less tedious. Nonetheless, as they exist today, applications in animation only serve to lessen rather than eliminate the labor intensiveness of accurate lip synching.

Laboratory products are being developed that will automate lip synching in animation in its entirety. In such an application all that is needed is the speech. The lip synch system converts the speech to parameters of both mouth shape and timing, and these are used to create the actual animation.

An **avatar** in computer-aided instruction is the embodiment of a teacher—usually a cartoon figure of some kind—that appears to be the source of instruction. A typical avatar might be a stick figure cartoon of a professorial nature with a pointer that highlights a pertinent part of the screen during a lesson. To add to the reality of the figure, its mouth moves as the computer lectures. Lip synch would make the mouth move realistically. Similarly, characters in computer games could be lip synched to add to their reality.

An application under discussion at North Carolina State University is a computer game in which the user provides the speech of one or more of the characters in the game, whose mouth moves in a realistic way through lip synching. This would provide the user with the additional amusement of making exorbitant sounds and seeing how the game character responds.

Virtual reality is the creation of an illusory world by computer into which a user is embedded via video, audio, and tactile impressions. Many virtual reality worlds are populated by human and animal figures who speak. Automated lip synch, as opposed to mindless opening and closing of the mouth, would contribute to the realistic effect of the virtual reality.

On the more serious side, automatic lip synching can be used as an aid to the hearing-impaired, especially for telephone communication. An automatic lip synching system would process speech coming over the telephone and animate a face on a screen as if it were speaking the speech, giving lip-reading aid to the hearing-impaired person. There would be a small time delay as the system processed the speech and then presented both speech and animation in synch, but that delay needn't exceed one or two seconds. Callers to the hearing-impaired person would be warned automatically that such a system is operating and to expect a somewhat different talking protocol than the usual one.

A similar system could be put in place to aid hearing in a noisy environment such as a factory floor. As observed in Chapter 5, even for persons with unimpaired hearing, the ability to see the mouth move during speech aids understanding.

All of the lip synching applications discussed so far are concerned with the external appearance of the mouth. That includes lip and jaw movement, and the occasional appearance of the tongue, as in the pronunciation of the phoneme /th/. Current research is also attempting to track the position of the entire tongue, whether visible from the outside or not. If this is done, lip synching can be applied to speech pathology. Patients could be shown the position of their tongue during an articulation and compare it with a target position, providing a powerful form of biofeedback for correcting speech defects.

References

[1] Chomsky, N., *Language and Mind,* New York, NY: Harcourt Brace Jovanovich, 1968.

[2] Meisel, W., "Telephones Present a New Face," *Speech Technology: The Magazine of Applied Speech Technology,* June/July 1998, pp. 16–19.

[3] Chin, D. B., and D. R. Pisoni, *Alcohol and Speech,* London: Academic Press, 1997.

[4] Phipps, T. C., and King, R. A., "Speaker Verification Biometric in Less Than 40 Bytes," *Speech Technology: The Magazine of Applied Speech Technology,* October/November 1997, pp. 30–35.

Glossary

access   Refers to the physical entry of a person into a secured area, or the electronic entry into a secured computer or computer file.

acoustic phonetics   The study of the physical properties of the sound waves of speech that are pertinent to human language understanding.

active vocabulary   A subset of the vocabulary of a speech recognition system to which the recognizer is restricted in certain contexts.

adaptive coder   A waveform coder that changes its step size (the magnitude of the difference between samples represented by the stored value) dynamically, according to how rapidly the signal is changing.

adaptive delta pulse code modulation (Adaptive delta PCM, ADPCM)   An adaptive coder based on delta PCM that uses a table lookup procedure to change its step size (the magnitude of the difference between samples represented by the stored value).

adaptive quantizer   A quantizer that changes its range, and hence its resolution, dynamically, adjusting to the changing range of the magnitude of the signal being quantized.

affricate   A consonant in which the obstruction is a stop followed by a fricative, for example /ch/, which is /t/ followed closely by /sh/.

aliasing   Caused by undersampling, it results in the higher frequency components of a signal being falsely reconstructed as lower frequencies. (*See also* **Nyquist rate, sampling theorem.**)

allodiphone   Diphones formed from the allophones of a language.

allophones   The speech sounds of language. The concrete sounds that correspond to the various phonemes; for example, /t/ is a phoneme, and it is realized as the allophones[t], [tʰ], and [D], among others, depending on context.

alveolar   Describes the place of articulation at the alveolar ridge.

alveolar ridge   The bony ridge of the hard palate behind the upper front teeth.

amplitude of a sine wave   The maximum amplitude reached during one period.

analog   Describes a signal that has an amplitude for every possible time; a continuous signal.

analog-to-digital (A-to-D)   The process by which an analog or continuous signal is represented digitally or discretely.

analog-to-digital converter (A-to-D converter)   An electronic device that samples an analog signal and represents it in a digital format.

anticipatory coarticulation   The influence on the pronunciation of an allophone due to the sound that comes immediately after it, for example, the pronouncing of /t/ as /ch/ in *bet you* (betcha).

antiformants   A band of frequency ranges of lower amplitude than surrounding frequencies. They are prominent in nasal consonants.

approximant   A consonant in which the obstruction is formed by two articulators not close enough together to cause turbulence; for example, /w/.

Arpabet   A phonetic alphabet devised for research projects funded by the Advanced Research Projects Administration (ARPA) of the U.S. Department of Defense.

articulators   The parts of the vocal tract that work together to produce speech sounds.

articulatory phonetics   That branch of phonetics concerned with the production of speech sounds.

artificial neural net (ANN)   Also called simply neural net, it is a computational model consisting of numerous processing units, any one of which is capable only of simple arithmetic or logical operations, connected into a complex network in which a given unit accepts inputs from other units, and produces an output, which may be fed into yet more units.

aspiration   The audible puff of air that follows a voiceless stop consonant in which voicing is resumed some milliseconds *after* the release of the stop closure.

avatar   In computer-aided instruction, the embodiment of a teacher or other character that interacts with the student.

average filter compensation   A means of reducing the mismatch between speaker models caused when the training phrase takes place through one channel, but the operational phase takes place through a different channel.

average percent eliminated (APE)   A figure of merit for speaker recognition systems that is sensitive to the population size over which the speaker identification takes place.

back   Describes a vowel articulated with the tongue deep in the mouth, away from the teeth; for example, /o/.

band-pass filter   A filter that removes frequencies both above and below certain cut-off levels.

bidirectional microphone   A microphone that collects sound coming from two directions only.

bilabial   Describes the place of articulation at the two lips; for example, /b/.

biometric   A measure of individuality based on a biological characteristic. Familiar biometrics are fingerprints, retinal patterns, handwriting patterns, facial features, blood types, DNA, and of course speaker recognition.

branching factor   The size of the active vocabulary; the number of vocabulary items from which the recognizer must choose.

central   Describes a vowel articulated with the tongue between the front position and the back position; for example, /ŭ/.

cepstral   Refers to properties of the cepstrum.

cepstrum   An anagram of spectrum, it is a display of speech achieved by taking the logarithm of the frequency domain representation and applying the inverse discrete Fourier transform to it, yielding a quasi-time domain representation of the signal. (*See also* **quefrency**.)

closed set speaker identification   Determining who among a group of persons a speaker is by the computational processing of his or her voice and of the voices in the group, when it is assumed the unknown person is in the group.

coarticulation   The influence on the pronunciation of an allophone due to surrounding allophones; for example, the "k" sound in *kick* is pronounced somewhat differently from the "k" sounds in *kayak*.

coda   The consonants of a syllable that follow the nucleus; for example /ch/ in *each*.

codebook (VQ codebook)   A table of codebook entries, and corresponding codebook values, that are computed by vector quantization as an abbreviated representation of sets of vectors.

codebook entry   A number, often in the range of 0 to $2n - 1$ for some n (codebook size), that is part of a codebook. It is used to represent the codebook values.

codebook value   The entry in a VQ codebook that represents the "average" value of a cluster of vectors.

code-excited linear prediction (CELP) vocoder   A kind of LPC vocoder designed to preserve the quality of the speech at lower bit rates.

cohorts   In forensic speaker verification, persons whose voice characteristics are similar to those of a suspect by virtue of being of the same age, size, and sex and speaking the same dialect as the suspect.

compression   The process of reducing the number of bits used to represent information with little or no information loss.

computer speech recognition   The ability to take speech as input and produce a transcript of that speech as output.

concatenative synthesis   The electronic synthesis of speech that is based on the rearrangement of prerecorded human speech, contrasted with parametric coding, which is the purely electronic synthesis of speech.

confusion matrix   A table of speech recognition performance wherein substitution errors are tabulated.

connected (speech)   Speech spoken with each word pronounced fully, though without necessarily a pause between words; for example, *didyou* instead of *didja*.

consonants   Speech sounds articulated with an obstruction in the oral cavity that restricts the flow of air.

continuous (speech)   Speech spoken with the words "run together." For example, saying *didja* instead of *did you*.

continuously variable slope delta modulation (CVSD)   A type of delta modulation in which the step size (the difference between samples represented by one bit) is changed dynamically to accommodate a rapidly changing signal. It is an example of an adaptive coder. It avoids some of the error of slope overload distortion.

creaky   Describes a speech sound pronounced with a very low rate of vocal cord vibration, giving it a "rusty gate" sound. Such pronunciations are not phonemic in English but may occur naturally in speech, especially at the ends of phrases, and interfere with the accuracy of speech recognition sometimes called "glottal fry".

decibel   A unitless measure of relative intensity of a sound defined as 20 times the base 10 logarithm of the ratio of the sound pressure level of the sound in question, to a reference sound pressure level that corresponds to the faintest sound perceptible to young, healthy ears.

deletion error   An error in speech recognition in which one or more of the words spoken is not reported by the recognizer. The error is usually a result of the user speaking too softly, or of too low a gain setting on the recognizer.

delta modulation   A type of difference coding in which one bit is used to represent the differences between sample values; for example, 1 is stored if the sampled value is larger than the previous one, and 0 is stored if the sampled value is not larger than the previous one. During reconstruction of the signal, 1 is interpreted as an increase, and 0 is interpreted as a decrease, in the signal, possibly leading to granular noise or quieting noise.

delta PCM   An instance of difference coding in which the quantized differences between sample values are used to represent the signal.

demisyllable   Either zero, one or more consonants followed by the first half of a vowel sound, or the second half of a vowel sound followed by zero, one, or more consonants. The word *seem* consists of two demisyllables: [sē|] and [|ēm], where the vertical stroke | indicates the missing half vowel sound.

dental   Describes the place of articulation where the tongue tip is inserted between the upper and lower teeth; for example, /th/.

dialect   The way a certain group of people, often distinguished geographically, render a language. Everyone speaks a dialect. Many people speak several dialects of their language that they use in different settings. A language is comprised of several dialects, no one of which is *linguistically* superior to any other, though certain dialects are thought of as "standard" because they are widely used.

difference coding   A method of waveform coding in which differences between successive sample values rather than absolute values are used to represent the signal.

digital-to-analog   The process by which an analog or continuous signal is reconstructed from its digital or discrete representation.

diphone   The second half of one allophone (speech sound) together with the first half of another allophone.

diphthong   A complex vowel sound that begins with one vowel and gradually changes to another vowel; for example, /ai/ in *bite*.

discrete (speech)   Speech spoken with each word pronounced fully and with a pause of about one-third of a second between words.

discrete Fourier transform (DFT)   A mathematical method for converting a digitized time domain representation of a signal into a frequency domain representation.

discrete-word recognition   The process of speech recognition where the vocabulary of words and/or phrases is stored as individual templates, and the user must pause between vocabulary items when using the system. Also called isolated-word recognition.

dynamic time warping (DTW)   The process of adjusting the duration of an utterance to match the duration of the template with which it is being compared during speech recognition.

dysarthria   Neurologically disturbed speech.

electronic synthesis   Often used to refer to parametric coding, also called rule-based synthesis.

elision   Linguistic jargon for the process in which a sound is omitted from a word, for example the second "e" in a casual pronunciation of *general* (genral).

equal error rate   The situation that obtains in a speaker verification system where the average false rejection error rate and the average false acceptance error rate are equal. It may be used as a figure of merit.

ergotic   Describes a hidden Markov model in which all transitions between all states have a non-zero probability; a totally connected hidden Markov model.

expectation   The notion that we expect people to say certain things, and not to say certain things, in a given situation. In speech recognition, this fact may be exploited to create smaller active vocabularies or choose among several equally likely recognition choices.

false acceptance   An error in speaker verification in which an impostor is verified as legitimate.

false alarm   An error in word-spotting speech recognition in which the recognizer takes a nonvocabulary word to be a word in its vocabulary.

false reject   An error in word-spotting speech recognition in which the recognizer misses a word in its vocabulary.

false rejection   An error in speaker verification in which a valid person is not verified.

filter   A device, possibly electronic, possibly computer software, that takes a signal as input, and outputs a signal with certain frequency components removed. (*See also* **high-pass filter, low-pass filter, band-pass filter.**)

flap   A consonant in which the obstruction is made by the tongue briefly touching the alveolar ridge; for example the sound of /t/ in the natural pronunciation of *writer* in many dialects of American English.

formant   A band of frequency ranges or harmonics in the spectrum of a vowel sound that have higher amplitude than surrounding frequencies. Vowel sounds are characterized by the lowest three formants, called the first formant, the second formant, and the third formant.

frame size   The width in milliseconds of the sequence of time windows over which successive LPC predictor coefficients are computed. (*See also* **linear predictive coding (LPC)**).

frequency   The number of pressure variation changes per second of a sound. It is perceived as pitch.

frequency domain display   A spectrum, that is, the graphical representation of a speech signal in which the *x*-axis measures frequency and the *y*-axis measures amplitude, at a given time interval.

fricative   A consonant in which the obstruction is formed by two articulators that are very close, but not touching, producing a partially obstructed, turbulent airstream; for example /s/.

front   Describes a vowel articulated with the tongue forward in the mouth, toward the teeth; for example /e/.

fusion   An error in continuous speech recognition in which the recognizer interprets two or more words to be less words; for example, mistakenly reporting *the row* as *zero*.

glottalic   Describes the place of articulation at the glottis; for example, /h/.

glottis   The opening between the vocal cords.

goat   The term used for users of speech recognizers for whom the system does not work well, most likely persons whose pronunciations are relatively inconsistent over time. (*See also* **sheep.**)

grammar   A schema or set of rules for designating the syntatically well-formed sentences drawn from a vocabulary of words.

granular noise   A term applied to describe the error that results during delta modulation when the signal is unvarying, since reconstruction requires some increase or decrease in the signal value. (*See also* **quieting noise**.)

harmonic A frequency whose value is a multiple of the fundamental frequency of the sound.

harmonic analysis The means of representing a periodic function as the sum of periodic functions whose frequencies are multiples or harmonics of the fundamental frequency of the periodic function being represented.

hertz (Hz) Vibrations per second. Used to describe frequency.

hidden Markov model(HMM) A two-stage probabilistic process represented graphically by means of states, transitions between states, and outputs, that can be used to model speech. They are used for the purpose of speech recognition. It is called "hidden" because the sequence of states that leads to a recognition decision is not necessarily known due to the model's probabilistic nature.

high Describes a vowel articulated with the tongue high in the mouth; for example, /ē/.

high-mid Describes a vowel articulated with the tongue between the high position and the mid position; for example, /i/.

high-pass filter A filter that removes frequencies below a certain cut-off level.

impostor In speaker verification, a person falsely claiming to be in the population of legitimate users.

insertion error An error in speech recognition in which one or more words are reported falsely as having been spoken. The error is usually a result of speaker or ambient noise being mistaken for a word by the recognizer.

intensity The amplitude or size of the pressure variation changes of a sound. It is perceived as loudness.

International Phonetic Alphabet (IPA) A phonetic alphabet devised by linguists to express unambiguously every speech sound of every language known in the world.

intonation The overall pitch contour of a phrase or sentence.

inverse discrete Fourier transform (IDFT) A mathematical method for converting the frequency domain representation of a signal into a time domain representation.

isolated-word recognition The process of speech recognition where the vocabulary of words and/or phrases is stored as individual templates, and the user must pause between vocabulary items when using the system. Also called discrete-word recognition.

labio-dental Describes the place of articulation where the lower lip is raised toward the upper teeth; for example, /f/.

language models A description of the probabilistic distribution of words in a language; for example, the fact that after the word *multiply,* words such as *by* and *the* are more probable than words such as *dog* or *eleemosynary.* Also called statistical language models.

latency Response time, as used in psychological or psycholinguistic experiments.

lateral approximant An "l" sound that is articulated by raising the tongue tip to the alveolar ridge, while the sides of the tongue rest on the back teeth; for example, most pronunciations of /l/ in American English.

lax Describes a class of vowels generally pronounced with somewhat less length and stress than tense vowels. Syllables in English cannot end with a lax vowel; for example, [bi] is not a possible English word.

length The duration of a segment.

letters Symbols of an alphabet that represent speech sounds.

liftering An anagram of filtering, it is a process analogous to filtering that takes place in the quefrency domain.

linear predictive coding (LPC) A method of implementing a vocoder in which the waveform source is analyzed for predictor coefficients that characterize the original waveform and are the basis for speech production. (*See also* **LPC decoder.**)

lip synchronization The process in which a computer takes a speech signal as input and outputs parameters for animating the lips, jaw, and tongue of a face such that the movements correspond to the speech.

logarithmic quantizer A nonlinear quantizer whose levels are taken logarithmically.

Lombard effect The tendency of people to speak louder in the presence of ambient noise.

loudness The perceived intensity of a sound; the amplitude or size of the pressure variation changes.

low Describes a vowel articulated with the tongue low in the mouth; for example, /a/.

low-mid Describes a vowel articulated with the tongue between the low position and the mid position; for example, /e/.

low-pass filter A filter that removes frequencies above a certain cut-off level.

LPC decoder An electronic device capable of reproducing speech when given as input the predictor coefficients produced by linear predictive coding (LPC).

manner of articulation The relative position and activity of the articulators at the obstruction during the pronunciation of a consonant; for example, fricative.

meter The relative segment length, pitch, and stress of allophones and syllables. Also called rhythm.

mid Describes a vowel articulated with the tongue about halfway between the **high** position and the **low** position; for example, /ā/.

model (reference model) A computer representation, often of a statistical nature, of vocabulary items for speech recognition, of speakers for speaker recognition, and of languages for language identification. (*See also* **templates**.) Models are more complex representations than templates, although they are often more efficient to use in terms of computer storage.

morpheme The smallest unit of linguistic meaning or function; for example, *im* in *impossible, es* in *fetches*.

multipulse LPC vocoder A kind of LPC vocoder that employs a technique for raising the quality of speech by improving the decoding of the signal in the higher frequency ranges.

murmured Describes a speech sound pronounced with partial vocal cord vibration, giving it a "breathy" sound. Such pronunciations are not phonemic in English but may occur naturally in speech and interfere with the accuracy of speech recognition.

narrowband spectrogram A spectrogram that is computed using windows of relatively long duration, thus emphasizing precision in the frequency aspect of the representation at the price of accuracy in the time aspect of the representation.

nasal Describes a speech sound in which air is allowed to pass through the nasal cavity and out the nose; for example, /m/.

nasal stop A speech sound in which air is blocked in the oral cavity but allowed to flow out of the nasal cavity; for example, /n/.

neural net (artificial neural net [ANN])   A computational model consisting of numerous processing units, any one of which is capable only of simple arithmetic or logical operations, connected in a complex network in which a given unit accepts inputs from other units, and produces an output, which may be fed into yet more units.

noise-canceling microphone   A microphone designed with circuitry to cancel ambient sound.

nucleus   The vocalic center of a syllable.

Nyquist rate   A sampling rate that is twice the frequency of the highest frequency component in the signal. (*See also* **sampling theorem, aliasing.**)

omni-directional microphone   A microphone that collects sounds from all directions. (*See also* **unidirectional microphone** and **bidirectional microphone.**)

onset   The consonants of a syllable preceding the nucleus; for example, /tr/ in *tree.*

open set speaker identification   Determining who among a group of persons a speaker is by the computational processing of his or her voice and of the voices in the group, allowing for the possibility that the unknown person may not be in the group.

operational phase   Regarding speaker and speech recognition, when a speaker or a speaker's utterance is identified based on the enrollment of his or her voice in a training phase.

oral   Describes a speech sound in which the sole egress of air is through the oral cavity, and no air enters the nasal cavity; for example, /s/.

palatal   Describes the place of articulation at the hard palate.

palatalization   A phonological process that causes sounds followed by a palatal consonant such as /y/ to have an altered pronunciation with the place of articulation shifted toward the palatal region; for example, the final /d/ in *did* is pronounced like [j] in *did yet* [dijet] in casually spoken speech.

parameters of sampling   The sampling rate and the quantizer.

parametric coding   The purely electronic synthesis of speech, contrasted with concatenative synthesis, which is based on prerecorded human speech.

periodic function   A function, usually with time measured on the x-axis, that repeats itself over some time interval; for example, a sine wave. The number of repetitions per second is the frequency of the function; the time elapsed between repetitions is the period; the *y*-value for any *x*-value (point in time) is the amplitude.

perplexity   For all intents and purposes, the average branching factor, though technically its computation is more complex than a simple average.

phoneme-based recognition   More appropriately called *allophone* recognition, it is speech recognition in which the basic units to be recognized are the speech sounds of the language.

phonemes   The basic sound units of a language. They are theoretical constructs with a basis in the psychology of language. Phonemes are pronounced as allophones, which are the concrete sounds that correspond to the phoneme, for example, /t/ is a phoneme and it is realized as [t], [t^h], and [D], among others, depending on context.

phonetic alphabet   A system of letter symbols in which one symbol represents one speech sound.

phonetics   The science of human speech sounds.

phonological rules   The patterns that determine which allophones of a phoneme are pronounced. For example, at the beginning of a syllable with primary stress, the allophone of /t/ is [t^h].

pitch   The fundamental frequency of a syllable, as perceived by the listener.

pitch synchronous overlap-add   Usually referred to as PSOLA it is a technique for changing the pitch, segment length, and loudness (hence, the stress and intonation) of spoken units of concatenative speech synthesis after waveform reconstruction. The technique may be applied to waveforms both in the time domain, where it is called time domain PSOLA (TD-PSOLA), and the frequency domain, where it is known as frequency domain PSOLA (FD-PSOLA).

place of articulation   The location in the oral cavity where airflow is restricted during the pronunciation of a consonant; for example, palatal.

post-alveolar   Describes the place of articulation just behind the alveolar ridge.

post-processing   In speech recognition, the application of higher level knowledge such as grammatical rules to guide decision making; for example, the recognizer might assign *men walk* and *men walks* to the same spectral pattern, but only *men walk* is grammatical, so it is chosen as the recognized sequence.

predictor coefficients   The coefficients on which linear predictive coding (LPC) is based. They indicate how a sampled value of a speech signal at any moment in time depends on the previous sample values.

primary stress   The most prominent stress in a word; for example, the stress on the syllable *duce* in the word *introduce*.

prosodic features   The syllable-wide properties of pitch, length, and loudness. Stress is also considered a prosodic feature. Also called suprasegmental features.

Prosody   The combined effect of prosodic features or suprasegmental features.

PSOLA   Stands for pitch synchronous overlap-add. It is a technique for changing the pitch, segment length and loudness (hence, the stress and intonation) of spoken units of concatenative speech synthesis after waveform reconstruction. The technique may be applied to waveforms both in the time domain, where it is called time domain PSOLA (TD-PSOLA), and the frequency domain, where it is known as frequency domain PSOLA (FD-PSOLA).

pulse code modulation (PCM)   A waveform coder that stores quantized samples directly; for example, a PCM coder with a sampling rate of 8,000 samples per second with an 8-bit quantizer uses 64,000 bits of storage per second of signal.

Quantization   The process of representing a range of numbers with a fixed number of digits, hence a finite, discrete number of values, rounding to the nearest representable value where necessary.

quantization error   The error that results from rounding to the nearest representable value during quantization.

quantization levels   The discrete values that result from quantization; for example a 4-bit quantizer has 16 quantization levels.

quantization noise   The deviation from the original signal of a reconstructed, quantized signal caused by the quantization error.

quantizer   The device that performs quantization; the process of performing quantization; the number of bits used by the quantizer. (*See also* **resolution of a quantizer.**)

quantizer step size   The difference between quantization levels.

quefrency   An anagram of frequency, it is the name given to the y-axis value of the cepstrum.

quefrency domain   Refers to the cepstrum.

quieting noise   A term applied to describe the error that results during delta modulation when the signal amplitude is zero, since reconstruction requires some increase or decrease in the signal value. (*See also* **granular noise.**)

reference model (model)   A computer representation, often of a statistical nature, of vocabulary items for speech recognition; of speakers for speaker recognition; and of languages for language identification. (*See also* **templates.**) Models are more complex representations than templates, although they are often more efficient to use in terms of computer storage.

rejection error   An error in speech recognition in which a word in the recognizer's vocabulary is spoken, but the recognizer signals that it was a nonvocabulary word or noise.

relative intensity   The intensity of a sound as measured in decibels (dB). It is perceived as loudness.

residual excited linear prediction (RELP) vocoder   A kind of LPC vocoder in which a residual error is computed, then transmitted or stored, and used to improve the accuracy of the speech reconstructed by the LPC decoder.

resolution of a quantizer   The number of bits used by the quantizer; also, the amplitude range encompassed by 2 to the power of the number of bits used by the quantizer.

rhythm   The relative segment length, pitch, and stress of allophones and syllables. Also called meter.

rounded   Describes a vowel articulated with the lips pursed; for example, /u/.

rule-based synthesis   Parametric coding or electronic synthesis.

sampling period   The amount of time between each sample when digitizing an analog signal. (*See also* **sampling rate.**)

sampling rate   The number of points taken each second when digitizing an analog signal. (*See also* **sampling period.**)

sampling theorem   The principle that a periodic signal must be sampled at greater than twice the Nyquist rate of its highest frequency component so that the signal may be accurately reconstructed from its digital representation. (*See also* **aliasing.**)

scrub   The act of editing electronic mail to be suitable for reading by a text-to-speech driven speech synthesizer.

secondary stress   The second most prominent stress in a word; for example, the stress on the syllable *in* the word *introduce*.

segment or phonetic segment   A term used to refer to a single speech sound in a syllable or word.

sheep   The term used for users of speech recognizers who are able to make the system work well, most likely persons whose pronunciations are relatively consistent over time. (*See also* **goat**.)

signal-to-noise ratio (SNR)   A measure in decibels of the amount of noise in a signal.

slope overload   A condition that occurs during delta modulation when the signal changes too rapidly to be faithfully represented by one bit. The condition results in distortion when the signal is reconstructed.

soft palate   Also called the velum, it is the back part of the roof of the mouth behind the hard palate. It is capable of moving to close off the nasal cavity during the articulation of oral (nonnasal) speech sounds.

sounds   Also called speech sounds, they are acoustical phenomena, to be distinguished from letters of the alphabet, which are *symbols representing* speech sounds. All human languages are built on a foundation of speech sounds, which may be combined into words and morphemes, which themselves may be combined into phrases and sentences capable of expressing any thought a speaker may have.

source coder   A vocoder; an electronic model of the vocal tract capable of producing speech when provided with input from the wave form of a speech signal, which is the "source" of the source coder.

speaker identification   Determining who a speaker is by the computational processing of his or her voice.

speaker recognition   The computational process of identifying or authenticating an individual based on his or her voice.

speaker verification   Determining that a person is who he says he is, based on the computational processing of his voice.

spectrogram   A graphical display of sound in which time is on the x-axis, frequency is on the y-axis, and the amplitude at a given time and frequency is indicated on a gray scale representation, where the darker the display, the higher the amplitude.

spectrographic analysis   The decomposing of sound into its component frequencies.

spectrum   The graphical representation of a speech signal in which the x-axis measures frequency and the y-axis measures amplitude, at a given time interval, also called a frequency domain display.

speech sounds　The individual sounds that are combined to make the syllables and words of a human language; for example, /i/, /p/.

speech synthesis　The electronic and computational process of producing
speech.

split　An error in continuous speech recognition in which the recognizer takes
one or more words to be more words; for example, *zero* as *the row*.

statistical language models　A description of the probabilistic distribution of
words in a language; for example, the fact that after the word *multiply* words
such as *by* and *the* are more probable than words such as *dog* or *eleemosynary*.
Also called language models.

stop consonant　Also called a stop, it is a consonant in which the obstruction
completely blocks the passage of air for several dozen milliseconds; for example, /t/.

stress　A constellation of properties of a syllable consisting of somewhat
greater length, loudness, and higher pitch.

style　The way a language is rendered as influenced by the social setting of the
speaker.

subband coding (SBC)　A type of waveform coding in which band-pass filters
are used to partition the signal into frequency subbands each of which is encoded separately, usually by PCM or ADPCM.

sublanguage　A collection of vocabulary words and rules for combining them
less general than the vocabulary and rules of the language as a whole.

subsetting　The process of restricting the overall vocabulary to the active vocabulary, depending on context; also called syntaxing.

substitution error　An error in speech recognition in which a certain word is
spoken, and a different word is reported by the recognizer.

suprasegmental (features)　Refers to properties that may extend over more
than one segment, often a syllable, but possibly an entire sentence, as when intonation is considered. The term is often used in reference to prosodic features.

syllabary　A writing system in which the symbols represent syllables.

syllable　The smallest utterable unit of speech, consisting of an onset, a nucleus, and a coda, of which only the nucleus is required. For example, the syllable /trim/ has as onset /tr/, as nucleus /i/, and as coda /m/; the final syllable
of *busy* is /ē/, consisting of a nucleus alone.

syntaxing The process of restricting the overall vocabulary to the active vocabulary, depending on context; also called subsetting.

template A computer representation of a vocabulary item in discrete word speech recognition.

tense Describes a class of vowels generally pronounced with somewhat more length and stress than lax vowels and capable of occurring at the ends of syllables in English; for example, /bē/ (*bee*).

text-dependent Speaker recognition based on the utterances of prescribed words. (*See also* **text-independent.**)

text-independent Speaker recognition based on any utterances, not necessarily prescribed in advance. (*See also* **text-dependent.**)

text-prompted speaker recognition A hybrid of text-independent speaker recognition and speech recognition, wherein both the speaker must be identified and the utterance must be recognized.

time domain display The graphical representation of a speech signal in which the *x*-axis measures time and the *y*-axis measures amplitude.

tone language A language where the pitch on the syllables in a word is part of what determines the meaning of the word; for example, Thai.

training The process of creating templates or models of vocabulary items for speech recognition, of speakers for speaker recognition, and of languages for language identification.

training phase Regarding speaker and speech recognition, when a speaker or a speaker's utterance is registered or enrolled for the purpose of later identifying that speaker by computer (speaker recognition), or identifying utterances of that speaker by computer (speech recognition). (*See also* **operational phase.**)

transition network A diagrammatical scheme for depicting the rules of a grammar using states and arcs.

trigram A triplet of words, used in creating statistical language models.

undersampling The condition obtained when the sampling rate is too low to allow the accurate reconstruction of the original signal. It may result in aliasing.

unidirectional microphone A microphone that collects sound coming from one direction only.

universal design The concept of applying technology in the design of a product such that the product is both assistive for impaired persons and useful for unimpaired persons; for example, the curb cuts one finds that allow a smooth transition between sidewalk and street, helpful to both persons in wheelchairs and persons pushing perambulators.

unrounded   Describes a vowel articulated with the lips spread; for example, /i/.

user's model   Facts about the performance of an individual user of a speech recognition system idiosyncratic to that user and that may be exploited to improve the performance of the system by that user.

uvula   The flap of tissue at the end of the velum.

vector quantization (VQ)   A method of compressing information contained in a set of vectors by taking advantage of clustering. Each cluster of vectors is represented by a single number, which is entered into a codebook and used for compression and transmission purposes.

vector sum excited linear predication (VSELP) vocoder   A kind of LPC vocoder designed to preserve the quality of the speech at lower bit rates.

velar   Describes the place of articulation at the velum or soft palate.

velum   Also called the soft palate, it is the back part of the roof of the mouth behind the hard palate. It is capable of moving to close off the nasal cavity during the articulation of oral (nonnasal) speech sounds.

virtual reality   The creation of an illusory world by computer into which a user is embedded via video, tactile, and audio impressions.

visemes   From _visual phoneme,_ the distinctive mouth shapes of language; for example, the lips-together position corresponding to a pronunciation of the phonemes /p/, /b/, and /m/ in English.

vocabulary   In speech recognition, the specific utterances for which the speech recognizer contains prestored templates or models.

vocoder   An electronic model of the vocal tract capable of producing speech when provided with input from the wave form of a speech signal; also called a source coder and voder.

voder   Another term for vocoder.

voice registration   The act of producing utterances during a training phase for speaker or speech recognition.

voiced   Describes a speech sound pronounced with vocal cord vibration; for example, /z/ as opposed to /s/.

voiceless Describes a speech sound pronounced without vocal cord vibration; for example, /s/ as opposed to /z/.

voiceprint The popular name used for a speech spectrogram.

voicing The state of the vocal cords during the articulation of a speech sound, which determines whether the sound is voiced or voiceless.

vowel A speech sound articulated with little or no obstruction in the vocal tract.

VQ codebook A table of codebook entries, and corresponding codebook values, that are computed by vector quantization as an abbreviated representation of sets of vectors.

waveform coding The process of representing the analog speech signal by sampling and quantizing.

waveform decoding The process of reconstructing a signal after it has been subjected to waveform coding.

wideband spectrogram A spectrogram that is computed using windows of relatively short duration, thus emphasizing precision in the time aspect of the representation at the price of accuracy in the frequency aspect of the representation.

windows When used in speech processing, successive intervals of time over which some process is carried out such as the computation of LPC coefficients.

word lattice Varying combinations of words that fit a spectral pattern of reference templates.

word-spotting A kind of speech recognition in which the words of the vocabulary are sought from a continuous flow of speech.

zero crossing The point at which a function's value is zero; for example, a sine wave has two zero crossings per period.

About the Author

Dr. Robert D. Rodman received a Masters degree in mathematics and a Ph.D. in linguistics at UCLA. He is currently on the faculty of the Computer Science Department at North Carolina State University. He was previously on the faculty at the University of North Carolina at Chapel Hill, and prior to that worked for General Electric Company, Burroughs Corporation and IBM. He has been publishing since 1963, including books on linguistics and speech processing.

Index

For further information on these and other Artech House titles,
including previously considered out-of-print books now available
through our In-Print-Forever™ (IPF™) program, contact:

Artech House	Artech House
685 Canton Street	46 Gillingham Street
Norwood, MA 02062	London SW1V 1Ah England
781-769-9750	+44 (0) 171-973-8077
Fax: 781-769-6334	Fax: +44 (0) 171-630-0166
Telex: 951-659	Telex: 951-659
email: artech@artech-house.com	email: artech-uk@artech-house.com

Find us on the World Wide Web at: www.artech-house.com